FLORA OF THE GUIANAS

Edited by

A.R.A. GÖRTS-VAN RIJN
&
M.J. JANSEN-JACOBS

Series A: Phanerogams
Fascicle 19

129. ANACARDIACEAE
(J.D. Mitchell)

including
Wood and Timber
(B.J.H. ter Welle, P. Détienne & T. Terrazas)

T0141415

1997

Royal Botanic Gardens, Kew

Contents

129. ANACARDIACEAE

by

JOHN D. MITCHELL[1]

Trees, shrubs, rarely subshrubs or vines, frequently with poisonous sap. Leaves alternate, rarely opposite or verticillate, simple or pinnately compound, petiolate or sessile; stipules absent; leaflets opposite or alternate, margin entire, serrate, dentate or crenate. Inflorescences terminal and/or axillary, thyrsoid, paniculate, racemose, or spicate, rarely cauliflorous; bracts and bracteoles deciduous or persistent. Flowers actinomorphic, unisexual or bisexual (plants dioecious, monoecious, andromonoecious, polygamous, or hermaphroditic); hypanthium sometimes present; perianth usually 2-whorled, rarely 1-whorled or absent; calyx (3-)4- or 5-lobed, rarely bracteate or calyptriform, persistent or deciduous, sometimes accrescent in fruit; petals (3-)4 or 5(-8), rarely absent, aestivation imbricate or valvate, deciduous or persistent, rarely accrescent in fruit; androecium usually actinomorphic, rarely zygomorphic, stamens (1-)5-10(->100), in 1 or 2 whorls, rarely more, in some genera only 1 or 2 fertile, filaments free, rarely basally connate into a tube, anthers dorsi- or basifixed, usually longitudinally dehiscent, introrse; disk intra-staminal, extra-staminal or absent; gynoecium 1-carpellate or syncarpous and 2-12-carpellate, rudimentary or absent in staminate flowers, ovary usually superior, rarely inferior, 1-5(-12)-locular, ovules 1 per locule, apotropous, attached basally, apically or laterally, styles 1-5(-12), terminal or lateral. Fruits drupaceous, fleshy or dry, sometimes with a lateral wing, occasionally subtended by an enlarged, chartaceous calyx, or attached to a fleshy hypocarp; seeds 1-5(-12), endosperm scanty or absent; embryo curved or straight, cotyledons usually free, bilobed, equal, planoconvex or reniform.

Distribution: World-wide, mainly in the tropics and subtropics: approx. 75 genera and 600 species; in the Neotropics 32 genera (1 cultivated and introduced from Asia) and approx. 140 species; in the Guianas 7 indigenous genera with 16 species, 2 introduced genera with 1 species each, and 1 indigenous genus with 2 introduced species.

Tribes of the **Anacardiaceae** in the Guianas:
- **Anacardieae**: Trees, rarely shrubs or subshrubs. Leaves simple, alternate, or opposite and decussate. Androecium sometimes

[1] The New York Botanical Garden, Bronx, New York 10458-5126, U.S.A.
Drawings by Bobby Angel.

zygomorphic; stamens (1-)5-many, usually 5-10; disk extra-staminal or lacking; carpel 1 (probably derived from 3 fused carpels), locule 1, ovule pendulous from basal funicle, style single, frequently lateral. In the Guianas: *Anacardium*, *Mangifera* (introduced).

- **Rhoeae**: Trees or shrubs, rarely vines. Leaves compound or simple. Stamens variable in number, usually 5-10; disk intra-staminal; carpels 3, united, with 1 locule, ovule apically, basally, or laterally attached, styles 1-3, terminal or lateral. In the Guianas: *Astronium*, *Loxopterygium*, *Schinus* (introduced), *Thyrsodium*.

- **Spondiadeae**: Trees, rarely shrubs or subshrubs. Leaves usually compound, rarely simple. Stamens twice the number of petals; disk intra-staminal; carpels united, 1-12, usually 4-5, locules usually 1-5, ovules pendulous from an apical funicle, styles usually 4-5, terminal. Opercula sometimes present on endocarp (see Plate 5 H). In the Guianas: *Cyrtocarpa*, *Spondias*, *Tapirira*.

Economic uses: Some of the species furnish valuable timber: *Astronium lecointei* and *Loxopterygium sagotii*. Two introduced species of *Spondias* (*S. dulcis*, *S. purpurea*), and *Mangifera indica* are cultivated for their fruits. *Schinus terebinthifolia* is rarely cultivated as an ornamental.

LITERATURE

Barfod, A. 1987. Anacardiaceae. In G. Harling & L. Andersson, Flora of Ecuador 30(104): 11-49.

Blackwell, W.H. & C.H. Dodson. 1967. Anacardiaceae. In R.E. Woodson et al., Flora of Panama, Ann. Missouri Bot. Gard. 54(3): 350-379.

Bornstein, A.J. 1989. Anacardiaceae. In R.A. Howard, Flora of the Lesser Antilles 5(2): 93-104.

Engler, A. 1876. Anacardiaceae. In C.F.P. von Martius, Flora Brasiliensis 12(2): 367-418.

Engler, A. 1882. Anacardiaceae. In A. & C. De Candolle, Monographiae Phanerogamarum 4: 171-500.

Engler, A. 1892. Anacardiaceae. In A. Engler & K. Prantl, Die Natürlichen Pflanzenfamilien, 3(5): 138-178.

Jansen-Jacobs, M.J. 1976. Anacardiaceae. In J. Lanjouw & A.L. Stoffers, Flora of Suriname 2(2): 441-444.

Marchand, N.L. 1869. Révision du groupe des Anacardiacées. Paris.

Mitchell, J.D. 1990. The poisonous Anacardiaceae genera of the World. Advances Econ. Bot. 8: 103-129.

Nannenga, E.T. 1934. Anacardiaceae. In A. Pulle, Flora of Suriname 2(1): 132-145.

Polak, A.M. 1992. Major timber trees of Guyana: a field guide. Tropenbos Series 2, Wageningen.

KEY TO THE GENERA (INCLUDING CULTIVATED ONES)

1 Leaves simple ·· 2
  Leaves compound ··· 3

2 Leaves usually obovate, apex obtuse to rounded, sometimes acuminate; stamens 8-10, 1 (or 2) fertile, much larger, filaments connate, forming a tube basally; disk absent; drupes with woody pericarp, on pedicel accrescent to a fleshy hypocarp ······························· *1. Anacardium*
  Leaves usually lanceolate, apex acute to acuminate; stamens 5, 1 or 2 fertile, larger, (this applies to the species cultivated in Neotropics), filaments free; extra-staminal disk present; drupes with fleshy mesocarp, hypocarp absent ··· *5. Mangifera*

3 Leaves and inflorescences exuding milky resin when broken; perianth forming a cupular hypanthium ····················· *9. Thyrsodium*
  Leaves and inflorescences without milky resin; hypanthium lacking ····· 4

4 Stamens 5; fruits dry ··· 5
  Stamens 8-10; fruits fleshy ··································· 6

5 Calyx of pistillate flowers enlarging several times their original size after anthesis; fruits without a lateral wing ················· *2. Astronium*
  Calyx of pistillate flowers not enlarging after anthesis; fruits with a lateral, falcate wing ······························ *4. Loxopterygium*

6 Leaflets with an intra-marginal vein; drupes 4- or 5-locular ··· *7. Spondias*
  Leaflets without an intra-marginal vein; drupes 1(-3)-locular ·········· 7

7 Leaf rachis frequently alate, secondary veins impressed to flattened abaxially; style 1 with 3 stigmas (this applies only to species cultivated in the Guianas); drupes pink to red at maturity, less than 7 mm long ········· *6. Schinus*
  Leaf rachis never alate, secondary veins prominent abaxially; styles 4-5, free; drupes reddish-purple to black at maturity, greater than 10 mm long ··· 8

8 Leaves deciduous; leaflets sessile or short-petiolulate; pistil glabrous; endocarp with an operculum ························· *3. Cyrtocarpa*
  Leaves evergreen; leaflets petiolulate; pistil densely pubescent; endocarp without an operculum ······························ *8. Tapirira*

1. **ANACARDIUM** L., Sp. Pl. 383. 1753; Gen. Pl., ed. 1: 129. 1737; Gen. Pl., ed. 5: 180. 1754.

Type: A. occidentale L.

Subshrubs to very large trees. Leaves aggregated toward branch tips, evergreen, alternate, simple, sessile to petiolate; blades chartaceous to coriaceous, oblanceolate to broadly obovate, sometimes ovate, oblong or elliptic, apex rounded, acute to obtuse, acuminate, emarginate, retuse or truncate, base cuneate, obtuse, attenuate or auriculate, margin entire; venation brochidodromous, domatia present in secondary vein axils abaxially. Inflorescences terminal and/or axillary, thyrsoid; lower bracts leaf-like, persistent or deciduous. Flowers pedicellate, bisexual and staminate (plants andromonoecious); calyx imbricate, 5-lobed; petals 5, imbricate, lower parts campanulate or forming a cylinder, tips reflexed; disk absent; stamens (6-)8-10(-12), with 1(-2) fertile, much larger, exserted on subulate or linear filaments, remaining stamens shorter and of variable length, their filaments connate basally to form staminal tube of variable length, anthers present or absent, thecae differentiated or undifferentiated and globose; pistil rudimentary in staminate flowers, ovary subglobose to obovoid, 1-locular, with 1 basal ovule, style terminal or lateral, stigma usually punctiform. Pedicels enlarge after fertilization to form obconical, pyriform or sigmoid hypocarp in all but 1 (extra-Guianan) species. Drupes reniform, or subreniform, pericarp woody with large rectangular cavities containing toxic oils; embryo curved, cotyledons plano-convex.

Distribution: Indigenous from Honduras, S to Paraguay and SE Brazil, and W of the Andes only in Venezuela, Colombia and Ecuador: 11 species, one of which is cultivated throughout the tropics; in the Guianas 5 species.

Literature: Mitchell, J.D. & S.A. Mori. 1987. The Cashew and its relatives (Anacardium: Anacardiaceae). Mem. New York Bot. Gard. 42: 1-76.

KEY TO THE SPECIES

1 Sepals less than 3 mm long; corolla campanulate, petals lanceolate, oblong or ovate; style usually lateral · 2
  Sepals more than 3 mm long; corolla cylindrical, petals linear to lorate; style terminal or subterminal · 4

2 Inflorescences less than 12 cm long; smaller stamens with normal thecae · 3
  Inflorescences more than 15 cm long; smaller stamens without visible thecae, their rudimentary anthers usually globose · · · · · · · · · · · · *3. A. giganteum*

3 Leaves usually chartaceous; petioles slender; tree to 40 m tall of rain forest
· *1. A. amapaënse*
Leaves coriaceous; petioles stout; shrub to small tree, 2-6 m tall of savannas
and granite outcrops · *2. A. fruticosum*

4 Basal bracts of inflorescence cream-colored to pale green adaxially; staminal
tube shorter than 2 mm; mature hypocarp yellow, orange, or red, never
white; small to medium-sized tree in open habitats or cultivated · · · · · · · ·
· *4. A. occidentale*
Basal bracts of inflorescence bright white adaxially and green abaxially;
staminal tube usually longer than 2 mm; mature hypocarp often white,
sometimes red or yellow; tall rainforest tree · · · · · · · · *5. A. spruceanum*

1. **Anacardium amapaënse** J.D. Mitch., Brittonia 44: 331. 1992. Type: Brazil, Amapá, R. Araguari, vic. Pedra Fina Camp and between Camps 12 and 13, Pires et al. 51650 (holotype MG, isotypes NY, RB, UB, US). – Plate 1 A-D.

Tree, to 40 m x 100 cm, with cylindrical trunk, unbuttressed to slightly buttressed at base, bark longitudinally fissured, inner bark resinous, trichomes sand-colored to golden, erect or appressed, to 0.1 mm long. Leaves glabrous; petiole slender, 6-24 mm long, sparsely pubescent; blade chartaceous to subcoriaceous, oblanceolate to obovate, 7.5-18.7 x 4-7 cm, apex short to long acuminate, occasionally rounded, base cuneate, attenuate or obtuse; primary vein flat adaxially, very prominent abaxially, secondary veins in 9-15 pairs, prominulous or impressed adaxially, prominent abaxially; domatia deep, pit-like. Inflorescences 5.8-6.5 cm long, sparsely to densely pubescent; peduncle 1 cm long; basal bracts foliose, obovate, sparsely pubescent or glabrous abaxially, distal bracts sepal-like, deciduous; pedicels densely pubescent, 1.5-2 mm long. Bisexual flowers: sepals narrowly ovate, 1.0-2.2 × 0.7-1.5 mm, pubescent abaxially; corolla campanulate, petals white or yellowish-green at anthesis, turning red after pollination, narrowly oblong, 4.7-6.7 x 1.1-1.6 mm, inrolled, densely pubescent adaxially, sparsely pubescent abaxially; staminal tube 1 mm long, slightly unequal in length around circumference, stamens 8-9, 1 much larger, with filament ca. 6 mm long, remaining filaments unequal in length, 1.0-3.0 mm long, with anthers of normal form; ovary 1.5 x 1.7-2 mm, glabrous, style slightly excentric, 5 mm long, stigma punctiform. Staminate flowers: similar to bisexual flowers in size of sepals and petals; pistillode 0.3 mm long. Fruit unknown.

D i s t r i b u t i o n : French Guiana and NE Amazonian Brazil (Amapá, E Pará and NW Maranhão); in evergreen, non-inundated high forest (FG: 1).

6

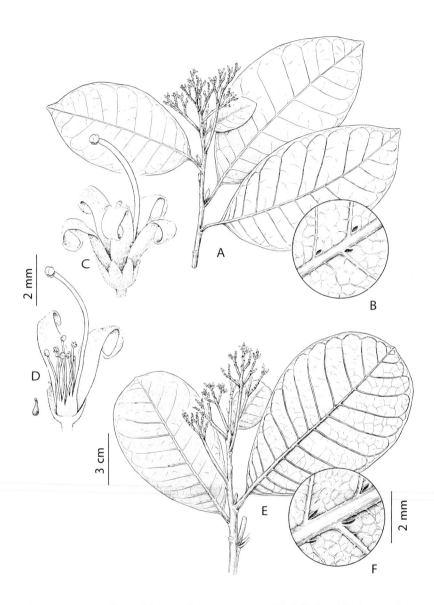

Plate 1. A comparison of *Anacardium amapaënse* J.D. Mitch. with *Anacardium fruticosum* J.D. Mitch. & S.A. Mori: A, habit; B, detail of abaxial leaf surface showing location of domatia; C-D, bisexual flowers of *A. amapaënse*; E, habit of *A. fruticosum*; F, close up of abaxial leaf surface showing location of domatia (A-B, Prance & Pennington 206; C-D, Pires et al. 51650; E-F, Maguire et al. 43865). Reprint from Mitchell, J.D. 1992. Brittonia 44(3): 331-338, fig. 3.

Specimen examined: French Guiana: Grand Inini R.-Basin of Maroni R., Sabatier & Prévost 3146 (CAY, NY).

2. **Anacardium fruticosum** J.D. Mitch. & S.A. Mori, Mem. New York Bot. Gard. 42: 54. 1987. Type: Guyana, Mazaruni-Potaro Distr., Karaurieng (Karowrieng in gazeteers) River Portage, 1250 m alt., Maguire & Fanshawe 32304 (holotype NY, isotypes MO, US).

– Plate 1 E-F.

Shrub to small tree, 2-6 m tall, trichomes whitish to beige-colored, erect or ascending to 0.1 mm long. Leaves glabrous; petiole stout, 10-13 mm long, sparsely pubescent; blade coriaceous, narrowly obovate to obovate, 13.5-15 x 7.7-8.7 cm, apex rounded or short-acuminate, base obtuse, cuneate or slightly attenuate; primary vein impressed adaxially, very prominent abaxially, secondary veins in 11-13 pairs, impressed adaxially, prominent abaxially; domatia deep, pit-like. Inflorescences 8.5-10.5 x 4.5-10 cm, pubescent; peduncle 10-15 mm long; basal bracts foliose, obovate, sparsely pubescent abaxially, distal bracts sepal-like deciduous; pedicels densely pubescent, 1.2-2 mm long. Bisexual flowers: sepals narrowly ovate, 2 x 1.1 mm, pubescent on both surfaces; corolla campanulate, petals white or pale green at anthesis, turning red after pollination, oblong to narrowly oblong, 4.5-5 x 1.6 mm, inrolled, densely pubescent on both surfaces; stamens 9, 1 much larger, with filament 4-5 mm long, remaining filaments unequal in length, 1.0-1.5 mm long, with anthers of normal form; ovary 1.2 x 1.3 mm, glabrous, style slightly excentric, 4 mm long, stigma punctiform. Staminate flowers: similar to bisexual flowers in size of sepals and petals; pistillode 0.3-0.8 mm long. Fruit unknown.

Distribution: Known only from the Upper Mazaruni R. Basin, Guyana; in savannas and on granite outcrops, alt. 460 to 1400 m (GU: 5).

Selected specimens: Guyana: Cuyuni-Mazaruni Region: Pakaraima Mts., 0.5 km NW of Imbaimadai settlement, 525-575 m alt., Hoffman 3416 (NY, US); near mouth of Partang R., Merume Mts., Upper Mazaruni R. Basin, 460 m alt., Maguire et al. 43865 (NY); Mazaruni-Potaro Region: Karowrieng R., unnamed Peak NW of Maipuri Falls, 1385 m alt., Pipoly & Alfred 7684 (NY, U, US).

3. **Anacardium giganteum** W. Hancock ex Engl. in Mart., Fl. Bras. 12(2): 409. 1876. Type: Brazil, Amazonas, R. Negro between Barcellos and S. Isabel, Spruce 1971 (holotype B, destroyed, isotypes BM, GH, K, M, NY, P).

Tree, to 40 m x 300 cm, with cylindrical trunk, unbuttressed, bark very thick, gray, moderately coarse, with vertical fissures, inner bark pinkish-brown, trichomes whitish to golden, appressed to erect, to 0.25 mm long. Petiole stout, 5-15 mm long, sparsely to densely pubescent; blade chartaceous to coriaceous, shiny adaxially, narrowly to broadly obovate, rarely oblanceolate, to 36.5 x 14 cm, apex usually rounded or obtuse, sometimes short-acuminate, emarginate or rarely truncate, base cuneate, obtuse, or slightly auriculate, adaxially glabrous, abaxially sparsely to densely pubescent on primary and secondary veins; primary vein very impressed to prominulous adaxially, prominent abaxially, secondary veins in 15-36 pairs, prominulous adaxially, prominent abaxially; domatia deep, pit-like. Inflorescences 16.5-26 x 17-29 cm, densely pubescent, especially toward distal branches of inflorescence; peduncle 0.5-2.5 cm long; basal bracts foliose, obovate, pubescent abaxially, distal bracts sepal-like, narrowly ovate; pedicels densely pubescent, 1.3-2 mm long. Bisexual flowers: sepals ovate, 1.4-2.3 x 1-1.5 mm, pubescent on both surfaces; corolla campanulate, petals yellow basally and white distally at anthesis, turning dark red after pollination, lanceolate to ovate, 4-5 x 1.5-2.2 mm, recurved, sparsely pubescent on both surfaces; staminal tube 0.5-1 mm long, slightly unequal in length around circumference, stamens 7-10, 1 much larger, with filament 3.8-4.5 mm long, remaining filaments ca. 1.0 mm long, rudimentary anthers with no apparent thecae, globose in form; ovary 1.5-2 x 1.2-2.2 mm, pubescent at apex, style excentric, 3-4.3 mm long, stigma punctiform. Staminate flowers: similar to bisexual flowers in size of sepals and petals; pistillode 1 mm long. Hypocarp at maturity pyriform, 1.5 x 1.3-3.5 cm, red. Fruits brown or black at maturity, subreniform, 2.5-2.7 x 1.8-2.5 cm.

Distribution: Colombia, Peru (E of the Andes), Bolivia (Amazonian frontier), Venezuela (Guayanan region), Amazonian Brazil (Acre, Amazonas, Maranhão, N Mato Grosso, Pará and Roraima), Guyana and Suriname; in evergreen, non-inundated, high forest (GU: 12; SU: 14).

Selected specimens: Guyana: NW District, Bonasika Landing, Arawau R., Archer 2312 (US); Mazaruni-Potaro Region, Kaieteur Falls, Potaro R., de la Cruz 4504 (GH, NY). Suriname: Nickerie District, Fallawatra, Jiménez-Saa 1500 = LBB 14241 (U); Sipaliwini Savanna area on Brazilian frontier, 335 m alt., Oldenburger et al. 590 (U).

Phenology: Flowering from Apr to Jan and fruiting from Jul to Feb.

Vernacular (and trade) names: Guyana: hooboodie, ubudi (Arow.). Suriname: boskasjoe (Sur.).

Uses: Wood used in construction. Hypocarp or cashew apples are edible, the quality varies from very sweet and tasty to very sour.

4. **Anacardium occidentale** L., Sp. Pl. 383. 1753. Lectotype (Fawcett & Rendle, Fl. Jamaica 5: 6. 1926; Mitchell & Mori, Mem. New York Bot. Gard. 42: 46. 1987): Sri Lanka, without locality, Flora Zeylanica 165 in Hermann Herbarium 3: 50 (BM).

Small tree, 1.5 to 10(-15) m x 40 cm, with broad crown and tortuous branching, bark brown or gray, smooth with scattered lenticels to rough with longitudinal fissures, inner bark thick, pale pinkish-orange to reddish-brown, trichomes white, erect, 0.1-0.2 mm long. Leaves glabrous; petiole slender to stout, 3-25 mm long; blade chartaceous to subcoriaceous, narrowly to broadly obovate sometimes broadly oblong, occasionally ovate or elliptic, 6.9-24 x 3.4-11.8 cm, apex usually rounded or obtuse, sometimes shortly acuminate, shallowly emarginate or truncate, base cuneate or obtuse, occasionally attenuate or auriculate; primary vein impressed to prominulous adaxially, very prominent abaxially, secondary veins in 8-18 pairs, prominulous adaxially, prominent abaxially; domatia shallow, basin-like. Inflorescences 11-29 x 4.5-24.5 cm, sparsely to densely pubescent toward distal branches of inflorescence; peduncle 1-6 cm long; basal bracts foliose, obovate, often light green adaxially, distal bracts sepal-like, lanceolate to ovate; pedicels sparsely to densely pubescent, 2.3-5 mm long. Bisexual flowers: sepals lanceolate to narrowly ovate, 3-6.5 x 0.8-2.0 mm, pubescent abaxially; corolla cylindrical from base to limb, petals white or pale green, often with pink or red lines at anthesis turning dark red after pollination, linear to lorate, (7-)8-13 x 1-2 mm, reflexed apically, sparsely pubescent on both surfaces; staminal tube 0.3-0.9 mm long, slightly unequal in length around circumference, stamens 5(-6)-10(-12), 1, rarely 2 much larger, with filament(s) 7.3-11.5 mm long, remaining filaments 2-4 mm long, all filaments with normal anthers; ovary 1.5 x 2.7 mm, glabrous, style terminal, 5.7-8.5 mm long, stigma punctiform. Staminate flowers: similar to bisexual flowers in size of sepals and petals; pistillode 0.3-1 mm long. Hypocarp at maturity, pyriform, much larger in cultivated forms than in wild populations, 5-20 x 2-8 cm, yellow, orange or red. Fruits gray or brown at maturity, subreniform, 2-3.5 x 1-2 cm.

Distribution: Cultivated and adventive throughout the Old and New World tropics; natural distribution is unclear because of its long association with man. However, it is probably indigenous from N South America, E of the Andes, S to São Paulo, Brazil (GU: 43; SU: 34; FG: 31).

Selected specimens: Guyana: Savanna near Lethem, Cooper 400 (U); Demerara-Mahaica Region, along Linden-Soesdyke Highway, ± 16 km S of Georgetown and W of Swan Settlement, Pipoly et al. 9162 (NY, US). Suriname: Donderskamp, Wayombo R., Jonker-Verhoef & Jonker 463 (U); Charlesburg Rift, 3 km N of Paramaribo, Maguire & Stahel 22743 (NY). French Guiana: In open savanna W of Cayenne, Cowan & Maguire 38029 (NY); Maripasoula, Fleury 875 (CAY, NY).

Phenology: Flowering and fruiting throughout the year.

Vernacular names: Guyana: cashew (English), merehe (Arow.), oroi (Carib.). Suriname: kadjoe, kasjoe (Sur.), olojé koe (Oyana). French Guiana: acajou, acajou à pomme, anacardier (French), bouchi (Paramak.), pomme cajou (French, commercial name).

Uses: Cultivated in the Guianas primarily for its edible hypocarps or cashew apples. The roasted seeds "nuts" are eaten as a snack and used in the manufacture of candies, cakes and cashew butter. India, E Africa and Brazil are the major areas of cashew production.

5. **Anacardium spruceanum** Benth. ex Engl. in Mart., Fl. Bras. 12(2): 410. 1876. Type: Brazil, Amazonas, N-shore of Amazon R. at mouth of R. Negro, Spruce 1684 (holotype B, destroyed, lectotype K (Mitchell & Mori, Mem. New York Bot. Gard. 42: 46. 1987, isolectotypes BM, GH, M, NY, P).   – Plate 2.

Tree, to 35 m x 100 cm, with cylindrical trunk slightly swollen at base, bark smooth with scattered lenticels, inner bark reddish-brown, trichomes golden, erect, to 0.1 mm long. Petiole slender, 15-40 mm long, glabrous; blade chartaceous, obovate, 9-20 x 3.5-10 cm, apex usually rounded or obtuse, sometimes acuminate, mucronate, shallowly emarginate or truncate, base usually cuneate, occasionally obtuse, glabrous on both surfaces except for an occasionally pubescent midrib abaxially; primary vein impressed adaxially, prominent abaxially, secondary veins in 13-16 pairs, prominulous adaxially, prominent abaxially; domatia deep, pit-like. Inflorescences 5-20.5 x 3-20 cm, sparsely to densely pubescent toward distal branches of inflorescence; peduncle 0.5-4 cm long; basal bracts foliose, obovate, usually bright white or occasionally pinkish-white adaxially, distal bracts sepal-like, lanceolate to ovate, pubescent; pedicels sparsely to densely pubescent, 1.5-3 mm long. Bisexual flowers: sepals lanceolate to narrowly ovate, 3.6-6.5 x 1.2-2 mm, glabrous adaxially, pubescent abaxially; corolla cylindrical from base to limb, petals pink to dark purple after pollination, linear to lorate, 6-12 x 1-1.5 mm, reflexed

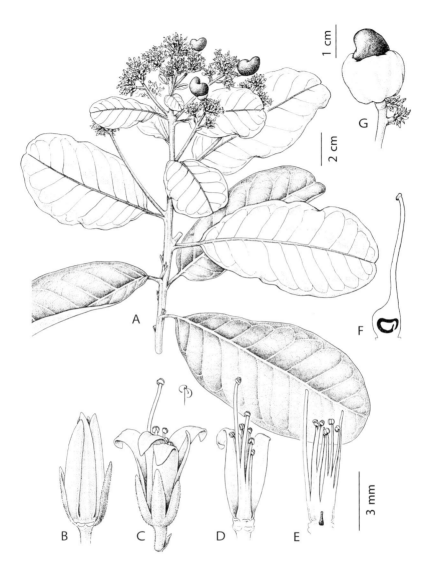

Plate 2. *Anacardium spruceanum* Benth. ex Engl.: A, habit, note the foliaceous bracts with the lighter adaxial surfaces; B, immature flower; C, staminate flower at anthesis; D, staminate flower with most of perianth removed; E, staminate flower with staminal tube slit open revealing varying degrees of filament coalescence, note the very small pistillode; F, pistil; G, fruit, in life the drupe is black and the hypocarp is either white or red (A-G, Mori et al. 15770). Reprint from Mitchell, J.D. & S.A. Mori. 1987. Mem. New York Bot. Gard. 42: 1-76, fig. 19.

apically, sparsely pubescent on both surfaces; staminal tube 1.2-3.5(-4.5) mm long, strongly unequal in length around circumference, stamens 8-10, 1 stamen much larger, with filament 5.2-7(-9) mm long, remaining filaments much shorter, with normal anthers; ovary 1.2-3 x 1.7-3.2 mm, glabrous, style usually terminal, stigma punctiform or capitate. Staminate flowers: similar to bisexual flowers in size of sepals and petals; pistillode 0.7-1 mm long. Hypocarp at maturity pyriform, 1 x 0.6-1.5 cm, white, red, yellow, with strong resinous smell. Fruits black at maturity, subreniform, 1.3-1.5 x 1.3-2 cm.

Distribution: Venezuela (Guayanan region), Amazonian Brazil (Amapá, Amazonas, Rondônia), Bolivia (Pando), Suriname, and French Guiana; in evergreen, non-inundated high forest and forest adjacent to rivers (SU: 6; FG: 14).

Selected specimens: Suriname: Nickerie, Fallawatra, Jiménez-Saa 1648 = LBB 14381 (NY, U); Brokopondo, 8 km ESE of Brownsweg village, van Donselaar 2041 (U). French Guiana: Saül, La Fumée Mt., 200-400 m alt., Mori, Pipoly & Mitchell 15684 (CAY, NY); Cr. Plomb, Basin of Sinnamary R., Loubry 1750 (CAY, NY).

Phenology: Flowering from Apr to Nov and fruiting from Jan to Dec.

Vernacular names: Suriname: boskasjoe (Sur.). French Guiana: bouchi kassoun (Paramak.).

2. **ASTRONIUM** Jacq., Enum. Syst. Pl. 10. 1760.
Type: A. graveolens Jacq.

Small to large trees. Leaves deciduous, alternate, imparipinnate; leaflets opposite, rarely alternate, petiolulate, margin entire to crenate, denticulate or serrate; venation prominulous to prominent, secondary veins not terminating at margin and freely ramified towards it. Inflorescences terminal or axillary, thyrsoid; bracts and bracteoles often deciduous; pedicels articulate. Flowers unisexual (plants dioecious); calyx and corolla imbricate, 5-merous; calyx enlarging after anthesis in pistillate flowers; disk intra-staminal, very thin, 5-lobate; stamens 5, alternating with petals and lobes of disk, filaments subulate, rudimentary or frequently absent in pistillate flowers; ovary 1-locular, with 1 apical ovule, styles 3, terminal, often persistent, stigmas capitate. Fruits fusiform, glabrous, subtended by enlarged, chartaceous calyx, mesocarp resinous, endocarp thin, brittle when dry; embryo straight, cotyledons plano-convex.

Distribution: Ranging from Mexico, throughout Central America, to Paraguay and N Argentina: 6-7 species; in the Guianas 3 species.

Literature: Barkley, F.A. 1968. Anacardiaceae: Rhoideae: Astronium. Phytologia 16(2): 107-152.
Mattick, F. 1934. Die Gattung Astronium, Notizbl. Bot. Gart. Berlin-Dahlem 11: 991-1012.
Santin, D.A. 1989. Revisão Taxonômica do Gênero Astronium Jacq. e Revalidação do Gênero Myracrodruon Fr. Allem. (Anacardiaceae). (Unpublished Master's Thesis, State University of Campinas, Campinas, São Paulo, Brazil).

KEY TO THE SPECIES

1 Leaflet blades glabrous to densely pubescent; distance between calyx and articulation of pedicel in fruit 1-3 mm · · · · · · · · · · · · *1. A. fraxinifolium*
Leaflet blades usually glabrous, occasionally sparsely pubescent on primary vein abaxially; distance between calyx and articulation of pedicel in fruit 5-15 mm · 2

2 Leaflets 7-15, usually 9, lanceolate, oblong or elliptic, apex long-acuminate, base strongly oblique · *2. A. lecointei*
Leaflets 3-7, usually 5, ovate, apex rounded to subacute, base slightly oblique · *3. A. ulei*

1. **Astronium fraxinifolium** Schott in Spreng., Syst. Veg., ed. 16, 4(2): 404. 1827. Neotype (Barkley, Phytologia 16(2): 125. 1968): Brazil, Bahia, Utinga, Blanchet 2765 (neotype G, isoneotypes BM, K, W).
– Plate 3 B.

Small to medium-sized tree, 5-20 x 4-40 cm, bark gray or brown, smooth on young trees becoming very rough with increasing age, exfoliating in large irregular patches, inner bark orange or reddish-brown with paler streaks, trichomes white, erect or crispate, to 0.5 mm long. Leaves 19.5-36.5 cm long, 7-13-foliolate; petiole glabrous to densely pubescent, 4.2-8 cm long; rachis glabrous to densely pubescent, 6-18 cm long; leaflets opposite or subopposite; petiolules sparsely to densely pubescent, lateral ones 2-6 mm long, terminal one 6-26 mm long; leaflet blades chartaceous to subcoriaceous, often lanceolate, less frequently oblong, elliptic or ovate, 7.1-12.7 x 2.7-7.1 cm, apex long-acuminate, base truncate, rounded or subcordate, sometimes slightly oblique, margin entire or occasionally denticulate, adaxially glabrous except primary vein sometimes sparsely to densely pubescent, abaxially sparsely to

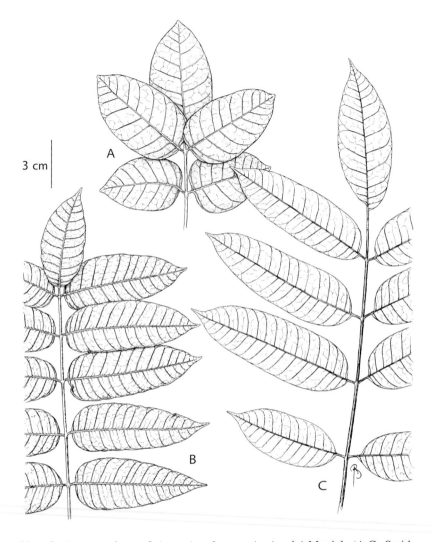

3 cm

A

B

C

Plate 3. A comparison of *Astronium* leaves: A, *A. ulei* Mattick (A.C. Smith 3466); B, *A. fraxinifolium* Schott (Oldenburger et al. 496); C, *A. lecointei* Ducke (Mori et al. 15806).

densely pubescent or glabrous; primary vein prominulous adaxially, prominent abaxially, secondary veins in 10-13 pairs, impressed to prominulous adaxially, prominent abaxially. Staminate inflorescences 12-25 cm long, glabrous to densely pubescent; peduncle 0-3 mm long; pedicels 1.2-1.7 mm long. Pistillate inflorescences and infructescences 17-37 cm long, glabrous to densely pubescent; peduncle 3-5 mm long;

pedicels 1-2 mm long. Staminate flowers: calyx lobes (sub)orbicular, 0.9-1.7 x 0.8-1.6 mm, glabrous; petals (greenish-)yellow to reddish, ovate to oblong, 2.4-3.5 x 1.2-1.9 mm, glabrous on both surfaces, apex obtuse; filaments 1.7-4 mm long, anthers oblong, 1.0-1.5 mm long. Pistillate flowers: calyx lobes suborbicular to broadly elliptic, 2.4-2.9 x 1.5-2 mm, glabrous; petals oblong to ovate, 1.6-1.8 x 0.8-1 mm, glabrous; ovary ovoid, 2-3.4 x 1.3-1.9 mm, glabrous, styles 0.3-0.4 mm long. Fruits fusiform, 10-14 x 3-4 mm, persistent and enlarged calyx lobes 10-15 mm long, distance between calyx base and articulation of pedicel in fruit 1-3 mm.

Distribution: Colombia, Peru, Bolivia, Argentina, Brazil S of Amazon R., Paraguay, S Suriname (Sipaliwini Savanna); in savanna (Suriname), chaco, caatinga, cerrados and tropical semi- to deciduous forests elsewhere, alt. 345 m in Suriname (SU: 1).

Specimen examined: Suriname: Sipaliwini Savanna area on Brazilian frontier, Oldenburger et al. 496 (NY, U).

2. **Astronium lecointei** Ducke, Arch. Jard. Bot. Rio de Janeiro 3: 202. 1922. Lectotype (Barkley, Phytologia 16(2): 118. 1968): Brazil, Pará, region of Trombetas R., E of lake Salgado, Ducke RB 13188 (lectotype RB, isolectotype K, P, U).   – Plate 3 C.

Tree, to 35 m x 62 cm, bark gray, smooth to rough, longitudinally fissured, inner bark cream-colored to light orange, trichomes white, crispate, to 0.1 mm long. Leaves 15-42 cm long, 7-15-foliolate; petiole glabrous, 3-9 cm long; rachis glabrous to sparsely pubescent, 5-18.7 cm long; leaflets usually opposite, occasionally subopposite; petiolules glabrous to sparsely pubescent, lateral ones 3-8 mm long, terminal one 7-32 mm long; leaflet blades chartaceous to coriaceous, shiny adaxially, usually narrowly oblong, less frequently lanceolate to narrowly ovate, narrowly obovate or elliptic, 5.1-15.5 x 2-4.6 cm, apex usually long-acuminate, acumen often mucronate, base truncate, rounded, obtuse or cuneate, usually strongly oblique, margin entire or with occasional minute teeth, rarely serrate, abaxially glabrous, or occasionally pubescent on primary vein; primary vein flat to prominulous adaxially, prominent abaxially, secondary veins in 8-7 pairs, flat to prominulous adaxially, prominulous to prominent abaxially. Staminate inflorescences to 28 cm long, glabrous; peduncle 0-1 cm long; pedicels 1-2.3 mm long. Pistillate inflorescences and infructescences to 35 cm long; peduncle 0-1 cm long; pedicels ca. 1.5 mm long. Staminate flowers: calyx lobes suborbicular, 0.9-1.2 x 1-1.3 mm, glabrous except for somewhat ciliate margin; petals greenish to cream-colored, ovate, 2.6-2.9 x 1.5-1.7 mm, apex obtuse, glabrous on both surfaces; filaments 2-2.8 mm long, anthers elliptic, 1.3-1.8 mm long. Pistillate flowers: calyx lobes suborbicular, 1.9 x 1.8 mm, glabrous except for sparsely ciliate margin; petals ovate, 2 x 1.2 mm, glabrous; rudimentary filaments ca. 0.7 mm long; ovary ovoid, 1.7 x 1.5 mm, glabrous, styles 0.3 mm long. Fruits fusiform, 9-11 x 4 mm, persistent and enlarged calyx lobes 9-12 mm long, distance between calyx base and articulation of pedicel in fruit 5-10 mm.

Distribution: Venezuela (mostly Guayanan region), Amazonian Brazil, E Ecuador, SE Peru, N Bolivia, Suriname; in semi-deciduous forest, evergreen, non-inundated high forest (SU: 4).

Selected specimens: Suriname: Jodensavanne-Mapane Cr. area, Suriname R., Lindeman 6856 (U); Near Camp 8, Mapane Cr. area, Schulz 7531 (U).

Uses: Valuable timber for general construction in Brazil.

3. **Astronium ulei** Mattick, Notizbl. Bot. Gart. Berlin-Dahlem 11: 996. 1934. Type: Brazil, Roraima, R. Branco, Ule 7959 (holotype B, destroyed, isotypes not seen). – Plates 3 A, 4.

Tree, to 40 m x 45 cm, buttresses 40-70 x 100-150 x 10-15 cm, concave near trunk, bark gray or dark-gray brown, somewhat reddish on buttresses, lenticels in irregular vertical rows, outer bark flaky on buttresses and adjacent trunk, inner bark light orange or light salmon-colored, with light brown layer, trichomes essentially lacking; exudate clear, somewhat sticky. Leaves 8-27.5 cm long, 3-7-foliolate; petiole glabrous, 2.5-7.5 cm long; rachis glabrous, 2.5-8 cm long; leaflets usually opposite, occasionally subopposite; petiolules glabrous, lateral ones 3-10 mm long, terminal one 13-22 mm long; leaflet blades chartaceous to subcoriacous, shiny adaxially, ovate, obovate or oblong, 4.5-12.5 x 2.2-6.6 cm, apex acuminate, acumen often mucronate, base truncate, rounded, obtuse or cuneate, symmetrical or slightly oblique, margin entire, on both surfaces glabrous; primary vein prominulous to prominent adaxially, prominent abaxially, secondary veins in 7-14 pairs, prominulous to prominent ad- and abaxially. Staminate inflorescences to 31 cm long, glabrous; peduncle to 1.5 cm long; pedicels 1.5-2 mm long. Pistillate inflorescences and infructescences 21-29 cm long; peduncle to 2 cm long; pedicels ca. 1.5 mm long. Staminate flowers: calyx lobes suborbicular, 0.5 x 0.5 mm, glabrous except for ciliate margin; petals greenish, ovate, 1.5-2 x 1 mm, apex obtuse; filaments 1.2-1.5 mm long, anthers oblong, 1 mm long. Pistillate flowers: calyx lobes orbicular, ca. 1.6 mm long, glabrous except for ciliate margin;

petals broadly ovate, 1.4-1.6 x 0.7 mm; ovary ovoid, 1 x 0.8-1 mm, glabrous, styles 0.5 mm long. Fruits fusiform, 1.1-13 x 4 mm, persistent and enlarged calyx lobes 10-15 mm long, distance between calyx base and articulation of pedicel in fruit 7-15 mm.

Distribution: Venezuela (Guayanan region), Brazil (Amazonas, Pará, Roraima), the Guianas; in semi-deciduous forest and evergreen, non-inundated high forest, alt. 0-400 m (GU: 5; SU: 3; FG: 5).

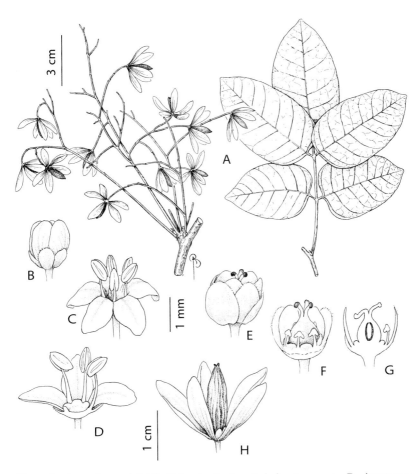

Plate 4. *Astronium ulei* Mattick: A, leaf and infructescence; B, immature staminate flower; C, staminate flower at anthesis; D, staminate flower with part of perianth removed; E, pistillate flower; F, pistillate flower with part of perianth removed; G, longitudinal section of pistillate flower showing position of ovule; H, fruit, note the enlarged sepals subtending the fruit (A & H, Rodriguez et al. 10237; B-D, Ule 7958; E-G, Steyermark 86578).

Selected specimens: Guyana: W extremity of Kanuku Mts., in drainage of Takutu R., 300 m alt., A.C. Smith 3126 (K, NY, P, U, W); Corantyne area, near branch of Epira R., concession of J. & Z. Sawh., 0-100 m alt., Polak 303 (NY, U). Suriname: Maratakka R., Helstone 227 (U); 1.5-2.0 km from base camp to Winanna Cr., tributary of Corantijn R., Maas & Tawjoeran in LBB 10734 (U). French Guiana: Road from Cayenne to St. Laurent, Prévost 2161 (CAY, NY, U); Nouragues Station, Basin of Arataye R., Sabatier & Prévost 1882 (CAY, NY).

Phenology: Flowering recorded in Aug and fruiting from Mar to Apr.

Vernacular names: Guyana: bauwaua (Wapisiana), bastard purpleheart (English).

3. **CYRTOCARPA** Kunth in Humb., Bonpl. & Kunth, Nov. Gen. Sp., Qu. ed. 7: 20, t. 609. 1824.
Type: C. procera Kunth

Small to medium-sized trees. Leaves deciduous, alternate, imparipinnate (rarely paripinnate), petiolate; leaflets opposite, occasionally subopposite, lateral leaflets sessile to short-petiolulate, margin entire, leaflets and rachis sericeous; venation eucamptodromous. Inflorescences subterminal, paniculate or pseudospicate. Flowers pedicellate, usually unisexual (plants dioecious or polygamodioecious); calyx and corolla imbricate, 5-merous; disk intra-staminal, annular, crenulate, fleshy; stamens 10, in 2 unequal series, outer (antesepalous) series longer; pistil glabrous; pistillode crowned by 3-5 stylodes; ovary 1-3(-5)-locular, ovules apical and only one developing, styles (3-)5, short, stigmas capitate. Drupes obliquely obtuse-oblong, styles often persistent, surface often pubescent, mesocarp fleshy, endocarp very thick, bony, 1-5-operculate; seed apparently 1, testa with saddle-shaped patch corresponding to hilum, cotyledons reniform.

Distribution: Mexico, Colombia, Venezuela, Brazil (Bahia, Roraima), Guyana: 4 species; 1 species in the Guianas.

Literature: Mitchell, J.D. & D.C. Daly. 1991. Cyrtocarpa Kunth (Anacardiaceae) in South America. Ann. Missouri Bot. Gard. 78: 184-189.

1. **Cyrtocarpa velutinifolia** (R.S. Cowan) J.D. Mitch. & Daly, Ann. Missouri Bot. Gard. 78: 186. 1991. – *Bursera velutinifolia* R.S. Cowan, Brittonia 7(5): 401. 1952. – *Tapirira velutinifolia* (R.S. Cowan) Marc.-Berti, Pittieria 13: 23. 1986. Type: Guyana, Sand Cr., Rupununi R., Wilson-Browne 112 (= FD 5650) (holotype NY, isotype K). – Plate 5.

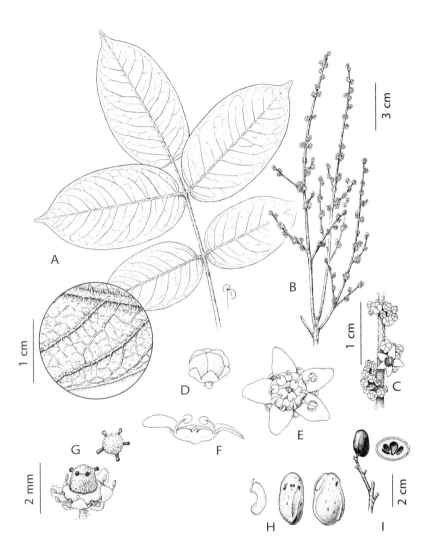

Plate 5. *Cyrtocarpa velutinifolia* (R.S. Cowan) J.D. Mitch. & Daly: A, part of mature leaf, including detail of abaxial leaflet surface; B, staminate inflorescence; C, detail of staminate inflorescence; D, bud; E, top view of staminate flower; F, longisection of staminate flower; G, pistillate flower, post-anthesis, including top view of pistil; H, embryo, dorsiventral and lateral views of endocarp (stone), showing operculum; I, portion of infructescence, transection of drupe (A, Bunting 4777; B-C, Goodland 1046; D-F, Wilson-Browne 518 = FD 5919; G & I, Wingfield 1011; H, Aristeguieta 6090). Reprinted from Mitchell, J.D. & D.G. Daly. 1991. Ann. Missouri Bot. Gard. 78(1): 184-189.

Loxopterygium gutierrezii Barkley, Lloydia 25: 121-122. 1962. Type: Colombia, Magdalena, 200 m elev., near Masinga, near Santa Marta, H.H. Smith 911 (holotype NY, isotypes A, BM, F, K, MICH, U, US).

Small tree, to 15 m x 41 cm, bark dark brown, flaky, longitudinally fissured, inner bark pink, laminated, trichomes whitish, ascending or erect, 0.6-0.9 mm. Leaves flushing after flowering, 18-35 cm long, 7-11-foliolate; petiole usually densely pubescent distally, 6-11 cm long; rachis usually densely pubescent, 7-16 cm long; leaflets usually opposite; petiolules densely pubescent, lateral ones 0-3 mm long, terminal one 10-25 mm long; leaflet blades chartaceous, oblong, elliptic, ovate or occasionally obovate, 4.5-10 x 2.9-5.5 cm, apex acuminate, base cordate or rounded, frequently oblique, on both surfaces sparsely to densely pubescent; primary vein impressed to prominulous adaxially, prominent abaxially, secondary veins in 8-12 pairs, impressed to prominulous adaxially, prominent abaxially. Inflorescences paniculate, secondary and tertiary axes pseudospicate, 11.5-31 cm long, densely pubescent; peduncle 0.3-0.5 cm long; bracts ovate to lanceolate, ca. 0.2 mm long; pedicels 0.5-0.8 mm long (staminate flowers) and ca. 1 mm long (pistillate flowers). Calyx lobes fleshy, deltate to broadly ovate, 0.9-0.1 x 0.8 mm, glabrous to sparsely pubescent abaxially; petals ovate, 1.8-2.0 x 0.9-1.1 mm, glabrous; disk thick, fleshy, 10-crenulate. Staminate flowers: filaments 0.8-0.9 mm (antepetalous) long and 1.0-1.1 mm (antesepalous) long, anthers 0.5 mm long; stylodes 0.3 mm long, glabrous. Pistillate flowers: rudimentary filaments 0.7 mm long (antepetalous) and 0.8 mm long (antesepalous), rudimentary anthers in dorsiventral view lanceolate, 0.3 mm long; ovary barrel-shaped with rounded apex, 1.1-1.5 x 1.3-1.7 mm, glabrous, locules usually 1 (rarely 2-5), ovule 1, styles 4-5, 0.4-0.5 mm long, separate, spreading. Drupes red to reddish-purple, obliquely oblong, 1.5-1.7 x 1.2 cm.

Distribution: NE Colombia, Venezuela, Brazil (Roraima), SW Guyana; in savannas, semi-deciduous forest, granite outcrops, alt. 100-300 m (GU: 7).

Selected specimens: Guyana: N Rupununi Savanna, Moreru [Moreiro in gazeteers], Goodland 1046 (NY); Dadanawa, Tawatawun Mt., Jansen-Jacobs et al. 2120 (U); Iramaikpang, Kanuku Mts., Wilson-Browne 518 (= FD 5919) (K, NY).

Phenology: Flowering from Sep to Nov and fruiting from Jan to Mar.

4. **LOXOPTERYGIUM** Hook. f. in Benth. & Hook., Gen. Pl. 1: 419. 1862.
Type: L. sagotii Hook. f.

Small to large trees. Leaves deciduous, alternate, imparipinnate; leaflets opposite or alternate, short-petiolulate, margin entire to crenate or serrate; secondary veins not terminating at margin, freely ramified towards it. Inflorescences axillary or rarely terminal, thyrsoid, consisting of alternating branches with densely congested cymes; pedicels articulate, or point of abscission inconspicuous. Flowers unisexual, rarely with bisexual flowers (plants dioecious or rarely polygamodioecious); calyx and corolla imbricate, 5-merous; disk intra-staminal, 5-lobed or subannular; stamens 5, inserted below disk, alternating with petals and lobes of disk, filaments subulate; pistil rudimentary in staminate flowers; ovary obovoid, compressed, unilocular with 1 basal ovule, styles 3, unequal, lateral. Fruits a falcate samara, lateral wing chartaceous with prominent venation, endocarp bony; embryo curved, cotyledons plano-convex.

Distribution: Venezuela (mostly NE Guayanan region), the Guianas, SW Ecuador, NW Peru, SW Bolivia, NW Argentina: 3 species; in the Guianas 1 species.

Literature: Barkley, F.A. 1962. Anacardiaceae: Rhoideae: Loxopterygium. Lloydia 25(2): 109-122.

1. **Loxopterygium sagotii** Hook. f. in Benth. & Hook., Gen. Pl. 1: 419. 1862. Type: French Guiana, Acarouany, 1857, Sagot 973 (lectotype P, here designated, isolectotypes BM, P, W). – Plate 6.

Tree, to 38 m x 73 cm, buttresses 40-100 x 10 cm, concave, slender, bark rough, gray or brown, inner bark laminated, with alternating layers of pale brown to pinkish brown with orange-brown, somewhat fibrous, trichomes whitish to golden, appressed, to 0.2 mm long; exudate white turning light green, somewhat sticky. Leaves 12-57.5 cm long, 5-10 foliolate; petiole sparsely pubescent, 2.4-14 cm long; rachis sparsely pubescent, 4-27 cm long; leaflets opposite or subopposite; petiolules pubescent, lateral ones 2-7 mm long, terminal one 9-35 mm long; leaflet blades chartaceous to subcoriaceous, usually narrowly oblong or lanceolate, sometimes elliptic, ovate or narrowly obovate, 5.2-15 x 2.2-5 cm, apex acuminate, base truncate, rounded or obtuse, slightly to strongly oblique, margin entire, adaxially glabrous, abaxially glabrous to sparsely pubescent; primary vein prominulous adaxially, prominent abaxially, secondary veins in 7-14 pairs, flattened or occasionally prominulous adaxially, prominulous to prominent abaxially. Inflorescences axillary, thyrsoid, 15-31 cm long, sparsely to densely pubescent; peduncle 0.8-6.6 cm long; bracts deltate, ovate or linear, 0.2-0.7 mm long, densely pubescent; pedicels to 0.7 mm long, densely

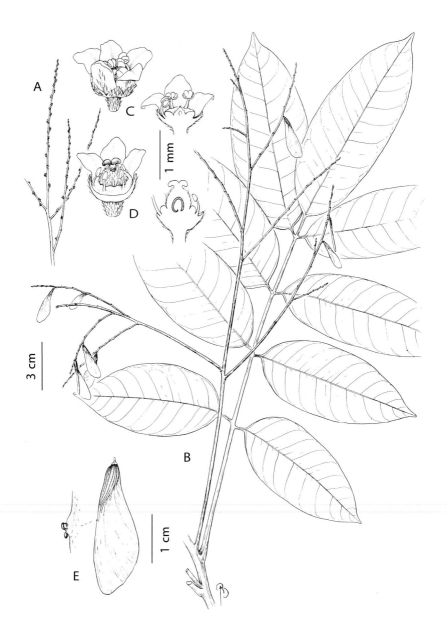

Plate 6. *Loxopterygium sagotii* Hook. f.: A, inflorescence; B, habit; C, staminate flower; D, pistillate flower; E, samara, note detail of persistent styles (3x) (A, Cox & Hubbard 82; D-E, Davidse & Gonzalez 16274; C, A.C. Smith 3512; D, BAFOG 7575).

pubescent. Flowers: calyx lobes deltate, 0.3-0.5 x 0.4-0.5 mm, densely pubescent abaxially; petals greenish, ovate, 0.8-1.0 x 0.6-0.8 mm, densely pubescent abaxially, primary vein black, prominent; disk fleshy, 5-lobed. Staminate flowers: filaments ca. 0.5 mm long, anthers ovate, 0.3 mm long; pistillode 0.2 mm long, densely pubescent. Pistillate flowers: rudimentary filaments 0.4-0.5 mm long, rudimentary anthers oblong or sagittate; ovary ovoid or obovoid, ca. 1 mm long, densely pubescent, styles unequal, very short, subapical, stigmas capitate or discoid. Samaras falcate, 2.3-3.4 cm long, calyx persistent but not enlarging, styles persistent on convex side of lateral wing, wing up to 1.5 cm wide.

Distribution: NE Guayanan Venezuela and the Guianas; in evergreen high forest, gallery forest and swamp forest, alt. 0-600 m (GU: 13; SU: 33; FG: 12).

Selected specimens: Guyana: Near Mazaruni Forest station, Archer 2434 (K); NW slopes of Kanuku Mts., in drainage of Moku-Moku Cr. (Takutu tributary), 150-400 m alt., A.C. Smith 3512 (F, NY, U). Suriname: Zanderij, Grounds of Lands Bosbeheer, Lems 5083 (NY); Upper Suriname R., near Goddo, Stahel 72 (K). French Guiana: 1 km S of Roche Koutou, Basin of Upper Marouini R., 250 m alt., de Granville et al. 9498 (CAY, NY, U); Mt. Mahury, near Cayenne, Mennega & Koek-Noorman 868 (U).

Phenology: Flowering recorded from every month of the year and fruiting from Aug to Jan.

Vernacular names: Guyana: hububalli (Arow.), kuipyari (Karib). Suriname: slangenhout, grootbladig slangenhout, kleinbladig slangenhout (Sur.), sneki-oedoe (Sran). French Guiana: koiha (Paramak.).

Uses: Timber for construction and furniture. The sap may induce a mild to severe contact dermatitis in sensitive individuals.

5. **MANGIFERA** L., Sp. Pl. 200. 1753.
   Type: M. indica L.

Trees. Leaves aggregated toward branch tips, evergreen, alternate, simple, petiolate; blades coriaceous, often lanceolate, apex acute to acuminate, margin entire. Inflorescences terminal or occasionally in upper leaf axils, pleiothyrsoid; pedicels articulate. Flowers bisexual and staminate (plants andromonoecious); calyx and corolla imbricate, 4- or 5-merous; petals often with glandular ridges adaxially; disk extra-staminal, 5-lobed, variously shaped; stamens 5(-12), 1 or 2 fertile, remaining stamens sterile, rarely 3-5 or all stamens fertile; pistil rudimentary in staminate flowers, ovary 1-locular with 1 basal ovule, style 1, lateral, stigma 1, punctiform. Drupes subglobose, subreniform or ovoid, mesocarp fleshy, usually thick, endocarp woody fibrous; embryo subreniform with plano-convex, often unequal and lobed cotyledons.

Distribution: Asian tropics, ca. 69 species, with *Mangifera indica* cultivated or naturalized throughout the worlds tropics and subtropics.

Literature: Kostermans, A.J.G.H. & J.M. Bompard. 1993. The Mangoes. Their Botany, Nomenclature, Horticulture and Utilization, Academic Press, London.
Mukherjee, S.K. 1985. Systematic and ecogeographic studies on crop genepools: 1. Mangifera L. International Board for Plant Genetic Resources, Rome.
Mukherji, S. 1949. A monograph on the genus Mangifera L., Lloydia 12(2): 73-136.

1. **Mangifera indica** L., Sp. Pl. 1: 200. 1753. Lectotype (Bornstein in Howard, Flora Lesser Antilles 5(2): 98. 1989): Rheede, Hort. Malab. 4: t. 1, 2. 1683.   – Plate 7.

Large tree, to 40 m x ca. 150 cm, bark grayish to blackish-brown, longitudinally fissured, inner bark yellow exuding yellowish-brown sap, trichomes present on inflorescences only, whitish to 0.2 mm long. Leaves glabrous; petiole 0.8-6 cm long; blades usually reddish when young, turning dark green and shiny when mature, thickly chartaceous to coriaceous, lanceolate, narrowly oblong or narrowly elliptic, 8-40 x 2-10 cm, apex acute to acuminate, base cuneate, shortly attenuate or obtuse; primary vein prominulous to prominent adaxially, prominent abaxially, secondary veins in 12-30 pairs, prominent on both surfaces, tertiary and higher order veins prominulous on both surfaces. Inflorescences 10-40(-60) cm long, sparsely to densely pubescent; peduncle 15-25 mm long; bracts lanceolate to ovate with acuminate apices, 15-25 mm long, glabrous to densely pubescent abaxially; pedicels 1-3 mm long, glabrous to densely pubescent. Bisexual flowers: perianth 5-merous, calyx lobes lanceolate to ovate, 2-2.6 x 1-1.5 mm, pubescent abaxially; petals initially greenish-white to cream-colored, later turning pink, elliptic to oblanceolate, 3-5 x 1-1.5 mm, apically reflexed, sparsely pubescent distally, 3-5 yellow to purple or brown ridges adaxially; disk of 5, often separate, very thick lobes; stamens 5, 1(-2) fertile with filament 3-5 mm long, remaining stamens sterile, 0.7-1 mm long; ovary depressed-globose, 1-1.5 mm long, style 1-2 mm long, lateral or excentric, curved, stigma punctiform. Drupes green, yellow, orange or red, variable in form and size, globose to oblong-ovoid or subreniform, sometimes laterally compressed, 8-30 cm long, mesocarp fleshy, orange, endocarp fibrous; seed laterally compressed.

Distribution: Cultivated and adventive throughout the Old and New World tropics, indigenous to the Asian tropics. (GU: 4; SU: 6; FG: 2).

Selected specimens: Guyana: Pomeroon Distr., Kabakaburi, de la Cruz 3342 (NY); NW Distr., Barima R., de la Cruz 3385 (NY). Suriname: Paramaribo, Indigenous Collector 181 (U); Kramer & Hekking 3140 (U). French Guiana: Maripasoula, Fleury 874 (CAY, NY); Saül, in village, ca. 200 m alt., Mori & Mitchell 18778 (NY).

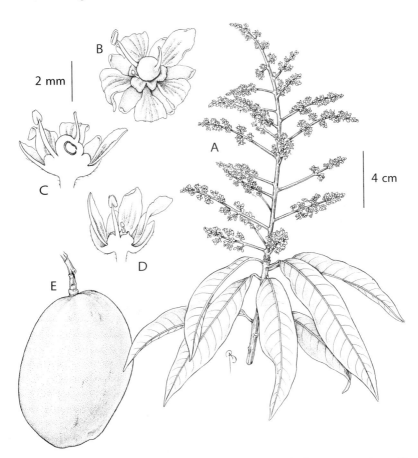

Plate 7. *Mangifera indica* L.: A, habit; B, bisexual flower top view; C, longisection of bisexual flower; D, longisection of staminate flower; E, fruit (A, Mori & Mitchell 18778; B-D, de La Cruz 3590; E, photo by author).

26

Phenology: Flowering Aug to Apr and fruiting period insufficiently documented by collections.

Uses: Cultivated in the Guianas for its edible fruit.

Vernacular names: Guyana: mango (English). Suriname: bobbie manja (Sur.). French Guiana: manguier (French), mã (Wayãpi, Palikur), pied manque (Créole).

6. **SCHINUS** L., Sp. Pl. 388. 1753; Gen. Pl., ed. 5: 184. 1754.
Lectotype (Y.J. Nasir in E. Nasir & S.I. Ali, Fl. Pakistan 152: 20. 1983): S. molle L.

Duvaua Kunth, Ann. Sci. Nat. (Paris) 2: 340. 1824.
Type: D. polygama Kunth = Schinus polygama (Cav.) Cabrera
Sarcotheca Turcz., Bjull. Moskovsk. Obsc. Isp. Prir., Otd. Biol. 1: 474. 1858, non Sarcotheca Blume (1850-51).
Type: S. bahiensis Turcz. = Schinus terebinthifolia Raddi

Trees or shrubs, sometimes thorny. Leaves evergreen or deciduous, alternate, imparipinnate, paripinnate or simple; leaflets opposite or alternate, usually sessile to subsessile, margin entire to crenate, denticulate or serrate. Inflorescences terminal or axillary, pleiothyrsoids; bracts small, deciduous; pedicels articulate. Flowers usually unisexual (plants usually dioecious); calyx and corolla imbricate, 5-merous; disk intra-staminal, crenulately 10-lobed; stamens (8-)10, filaments subulate; pistil rudimentary in staminate flowers; ovary 1-locular with 1 lateral to apical ovule, styles 3 or single, terminal, stigmas 3, capitate. Drupes small, globose, exocarp thin, shining, deciduous, mesocarp fleshy, resinous, endocarp bony; seed compressed, cotyledons flat.

Distribution: Ecuador, Peru, E and C Brazil, S to Patagonia: 20-24 species, 1 of which, S. terebinthifolia is rarely cultivated in Guyana.

Literature: Barkley, F.A. 1944. Schinus L. Brittonia 5(2): 160-198. Barkley, F.A. 1957. A study of Schinus L. Lilloa 28: 5-110.

1. **Schinus terebinthifolia** Raddi, Mem. Mat. Fis. Soc. Ital. Sci. Modena, Pt. Mem. Fis. 18: 399. 1820. Type: Brazil, without locality, Raddi s.n. (FI). – Plate 8.

Shrub to small tree, to 7 m tall, bark brown, longitudinally fissured, trichomes to 0.4 mm long, erect, whitish. Leaves usually imparipinnate, 7-22 cm long, (3-)5-15-foliolate; petiole glabrous to pubescent, 1.5-4 cm long; rachis glabrous to sparsely pubescent, 5-15 cm long, often alate; leaflets opposite, lateral ones sessile to subsessile, terminal one sessile with attenuate base or with petiole to 1 cm long; leaflet blades chartaceous to subcoriaceous, shiny adaxially, elliptic, oblong, obovate, lanceolate to ovate, 2.0-6 x 0.8-3 cm, apex acute to

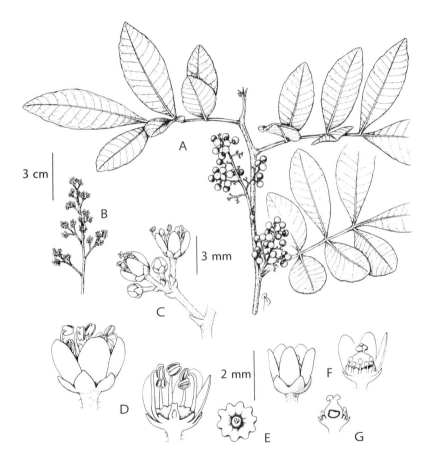

Plate 8. *Schinus terebinthifolia* Raddi: A, habit; B, staminate inflorescence; C, detail of cymose arrangement of staminate flowers; D, staminate flower (external view) and in longisection; E, top view of disk and pistillode; F, pistillate flower (external view) and with part of perianth removed; G, longisection of pistillate flower (A, habit based on field sketch by Bobbi Angell; B-E, Howard 9992; F-G, Zanoni 25786). Reprinted from Acevedo, P., 1996. Mem. New York Bot. Gard. 78: 1-581.

obtuse or rounded, occasionally mucronulate, base obtuse or rounded to cuneate or attenuate, symmetrical or slightly oblique, margin entire to serrulate-crenulate often revolute, both surfaces glabrous except primary vein sometimes sparsely pubescent; primary vein prominent adaxially, prominulous to flattened abaxially, secondary veins in 7-14 pairs, prominulous adaxially, impressed to flattened abaxially. Inflorescences 2-15 cm long, glabrous to sparsely pubescent; bracts lanceolate to deltate, 0.3-0.7 mm long, glabrous to sparsely pubescent abaxially, margin ciliate; peduncle 0.3-2.0 cm long; pedicels 0.5-3 mm long, sparsely pubescent below articulation. Flowers: calyx lobes deltate, 0.5-0.8 mm long, glabrous to sparsely pubescent, margin sometimes ciliate; petals white, lanceolate to obovate, 1.5-2.5 x 1-1.5 mm, apex obtuse, glabrous on both surfaces; disk crenulately 10-lobed. Staminate flowers: stamens 10, filaments ca. 0.4-1.0 mm (antepetalous) long and 1.0-2.0 mm (antesepalous) long, anthers ovoid to spherical, 0.5-0.7 mm long; pistillode a very reduced version of pistil, ca. 0.5 mm long. Pistillate flowers: staminodes with rudimentary anthers, 0.5-1.0 mm long; ovary subglobose, ca. 1 mm long, glabrous, style 1, 0.2-0.3 mm long, stigmas 3, capitate. Drupes pink or red, at maturity globose, ca. 0.5 cm in diam.

Distribution: Indigenous in S Brazil, Paraguay and NE Argentina; cultivated and naturalized in subtropical and tropical areas of both hemispheres; planted as an ornamental in Georgetown, Guyana (GU: 1).

Specimen examined: Guyana: W Demerara Region, Georgetown, Guyana Botanical Garden, 0-10 m alt., Pipoly 7346 (NY).

Uses: Planted as an ornamental for its foliage and pink to red fruits which are the source of "pink peppercorns" of fancy cuisine. Various parts of the plant induce a mild to severe contact dermatitis in sensitive individuals.

7. **SPONDIAS** L., Sp. Pl. 200. 1753.
   Type: S. mombin L.

Small to large trees. Leaves deciduous, alternate, aggregated toward branch tips, usually imparipinnate; leaflets (sub)opposite, petiolulate, entire, crenate, or serrate, intra-marginal vein present. Inflorescences terminal or axillary, paniculate; pedicels articulate. Flowers bisexual, strongly protandrous, 5-merous; calyx lobes imbricate or free; corolla valvate, petals cucullate; disk intra-staminal, annular, usually continuous, notched and undulate or crenulate; stamens (8-)10, in 2 (sometimes strongly) unequal series, filaments linear to subulate; ovary

(3-)5-locular, each locule with 1 apical ovule, styles (3-)5, stigmas capitate to oblique. Drupes oblong, globose, obovoid or ellipsoid, mesocarp fleshy, edible, sour, growing over apex of pedicel (*S. dulcis*), endocarp bony with fibrous matrix or with spiny projections; cotyledons linear, opposite, green, sessile, somewhat fleshy.

Distribution: Mexico to S Brazil in the New World, and in tropical Asia. Introduced and occasionally escaped throughout tropical Africa (*S. mombin* and *S. purpurea*) and to the West Indies, where the genus is probably not native. Approximately 18 species world wide; the Neotropics 8-9 species, 1 of which (*S. dulcis*) introduced from the S Pacific; in the Guianas 3 species, 2 of which (*S. dulcis* and *S. purpurea*) are introduced.

Literature: Airy Shaw, H.K. & L.L. Forman. 1967. The genus Spondias L. (Anacardiaceae) in tropical Asia. Kew Bull. 21(1): 1-19. Kostermans, A.J.G.H. 1991. Kedondong, Ambarella, Amra. The Spondiadeae (Anacardiaceae) in Asia and the Pacific Area (with some notes on introduced American species). Yayasan Tumbuh tum buhan yang berguna Julid I (Foundation Useful Plants of Asia Vol. I). Herbarium Bogoriense, Bogor, Indonesia.

KEY TO THE SPECIES

1 Inflorescences axillary, few-flowered; petals red to purple; lateral petiolules 0-3 mm long; leaflets 3-6 cm long; adaxial primary vein sparsely to densely pubescent · *3. S. purpurea.*
  Inflorescences terminal (occasionally arising from upper leaf axils), many-flowered; petals cream-colored to greenish-white; lateral petiolules 3-10 mm long; leaflets 5-15 cm long; adaxial primary vein glabrous · · · · · · · 2

2 Leaflets nearly symmetrical, margin usually serrulate; inflorescences appearing before or together with young leaves; fruits 4-10 cm long, mesocarp hiding apex of pedicel, endocarp with spiny projections · · · · · ·
 · *1. S. dulcis*
  Leaflets somewhat falcate, margin usually entire; inflorescences present together with mature leaves; fruits up to 4 cm long, mesocarp not hiding apex of pedicel, endocarp without spiny projections · · · · · · · *2. S. mombin*

1. **Spondias dulcis** Parkinson, J. Voy. South Seas 39. 1773; G. Forster, Pl. Esc. 33. 1786. Lectotype (A.C. Smith, Flora Vitiensis Nova 3: 452. 1985): Tahiti, without locality, "Capt. Cook" (BM).

Spondias cytherea Sonn., Voy. Indes. Orient. 3: 242. t. 123. 1782. Type: Tahiti, Commerson s.n. (holotype P).

Small to medium-sized tree, to 25 m x 40 cm, outer bark light gray or light brown, thin, smooth to moderately rough, trichomes essentially absent. Leaves 11-60 cm long, 9-25-foliolate; petiole glabrous, 9-15 cm long; rachis glabrous, 17-40 cm long; leaflets opposite (subopposite); petiolules glabrous, lateral ones 2-8 mm long, terminal one 1-3 cm long; leaflet blades chartaceous, oblong or lanceolate to ovate, nearly symmetrical, 5-15 x 1.7-5 cm long, apex acuminate or occasionally acute, base cuneate or obtuse, nearly equilateral, margin usually serrulate or crenulate, glabrous on both surfaces; primary vein prominulous adaxially, prominulous to prominent abaxially, secondary veins in 12-20 pairs, impressed to prominulous adaxially, flattened abaxially. Inflorescences terminal, 17-35 cm long, many-flowered, developing before or with young leaves, glabrous; peduncle 0-0.1 cm; bracts linear to lanceolate, 0.4-5 mm long, apex acute; pedicels 1-3 mm long, glabrous, articulated. Flowers with calyx lobes deltate, 0.5-1.0 mm long, glabrous; petals cream-colored or white, oblong or ovate, 2-3 x 1-1.3 mm, apically cucullate, glabrous; stamens (9-)10, antepetalous ones ca. 1.5 mm long, antesepalous ones 1.7-2.0 mm long, anthers oblong, 0.7-0.8 mm long; disk crenulate; pistil 2 x 1.8 mm, ovary depressed globose, 0.6-1.0 mm long, glabrous, styles ca. 0.8 mm long, stigmas oblique. Drupes yellow or orange at maturity, ellipsoid, obovoid, or oblong, 4-10 x 3-8 cm, mesocarp hiding apex of pedicel at maturity, endocarp bearing spiny processes.

Distribution: Indigenous to Polynesia; cultivated throughout the tropics of the world including the Guianas (GU: 1; FG: 2).

Specimens examined: Guyana: Diamond, E bank of Demerara R., Omawale & Persaud 94 (NY). French Guiana: Without locality, Sagot s.n. (P); Acarouany, Sagot 196 (P).

Vernacular names: Suriname: fransi mope (Sur.). French Guiana: pomme cythère, prune de cythère (Créole, French).

Uses: Cultivated for its edible fruits which are used to make juice or to flavor ice cream.

2. **Spondias mombin** L., Sp. Pl. 1: 371. 1753. Lectotype (Bornstein in Howard, Flora Lesser Antilles 5(2): 101. 1989): Merian, Metamorph. Insect. Surinam. t. 13. 1705. – *Spondias lutea* L., Sp. Pl., ed. 2: 613. 1762, nom. illegit. for Spondias mombin L

– Plate 9 A-D.

Medium-sized to large tree, to 25 m x 56 cm, outer bark brown or gray, smooth or often rough, with longitudinal fissures, sometimes with

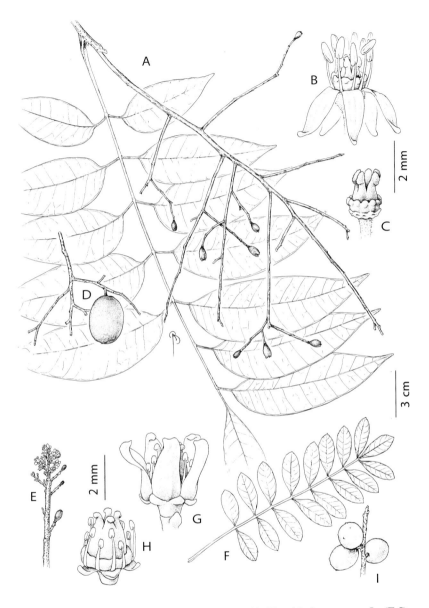

Plate 9. A comparison of *Spondias mombin* L. (A-D) with *S. purpurea* L. (E-I): A, habit; B, bisexual flower; C, bisexual flower with perianth and stamens removed; D, infructescence; E, inflorescence; F, leaf habit; G, bisexual flower; H, bisexual flower with petals removed; I, section of infructescence (A, Mori et al. 21534; B-C, de La Cruz 3839; D, Lanjouw & Lindeman 3152; E, G-H, Daly 137; F & I, Nee 37681).

spinose, corky outgrowths, shed in rectangular plates, inner bark pinkish-orange, trichomes white, erect to 0.2 mm long. Leaves 18-42.5 cm long, 7-15-foliolate; petiole 4-13.5 cm long, glabrous to pubescent; rachis 7.1-25 cm long, usually sparsely pubescent; leaflets (sub)opposite; petiolules pubescent or glabrous, lateral ones 3-10 mm long, terminal one 14-40 mm long; leaflet blades chartaceous or occasionally subcoriaceous, usually narrowly oblong, occasionally narrowly ovate, lanceolate or elliptic, somewhat falcate, 6-15 x 2.8-4.5 cm, apex acuminate or occasionally acute, base truncate or obtuse, oblique, margin (sub)entire (on seedlings first expanded leaflet blades crenate to serrate), adaxially glabrous, abaxially glabrous except primary vein and secondary veins sometimes pubescent, primary vein prominulous adaxially, prominent abaxially, secondary veins in 8-20 pairs, prominulous to slightly impressed adaxially, prominent abaxially. Inflorescences terminal and/or axillary, 15-60 cm long, many-flowered, developing when mature leaves are present, usually pubescent, occasionally glabrous; peduncle 1-10 cm long; secondary axes well-developed; bracts lanceolate to ovate, 0.4-1 mm long, apex acute or acuminate, pubescent abaxially; pedicels 2-4.5 mm long, pubescent, articulate. Flowers with calyx lobes deltate, 0.4-0.6 mm long, sparsely pubescent abaxially; petals cream-colored or white, lanceolate, 2.5-3.2 mm long, apically cucullate, glabrous; stamens 10, antepetalous ones 2-2.3 mm long, antesepalous ones 2.5-2.7 mm long, anthers oblong, 1 mm long; disk crenulate; pistil 1.3-1.6 mm long, ovary essentially ovoid, 0.6 mm long, glabrous, styles 0.7-1 mm long. Drupes yellow or orange, oblong or less often ellipsoid or slightly ovoid-oblong, 2-4 x 1.8-2.6 cm, endocarp oblong, bony.

Distribution: Cultivated and adventive sporadically in tropical Africa, Asia, and the West Indies; indigenous from S Mexico, S through Central America and N South America to Bolivia, and SE Brazil; in semi-deciduous, gallery and evergreen, non-inundated high forest, alt. 0-400 m (GU: 25; SU: 13; FG: 12).

Selected specimens: Guyana: Vicinity of Bartica, Essequibo R., de la Cruz 1951 (NY); Kanuku Mts., 140-300 m alt., Jansen-Jacobs et al. 734 (NY, U). Suriname: Nickerie, on small ridge S of Clara polder, Lanjouw & Lindeman 3152 (K, U); Paramaribo, Wullschlägel 930 (W). French Guiana: Centre ORSTOM, Ile de Cayenne, 4° 56'N, 52° 19'W, Bordenave 173 (CAY, NY); Saül, Along Route de Belizon at "army camp", 511 meters N of Eaux Claires, 200-400 m alt., Mori et al. 21534 (CAY, NY).

Phenology: Flowering from Aug to Apr and fruiting from Aug to May.

Vernacular names: Guyana: hog plum (English), hubu, mope (Carib.), plum tree (English). Suriname: mope. French Guiana: akaya

(Wayãpi of the upper Oyapock), caja (Portuguese), kaxambag (Palikur), mombin (Créole), mope (Galibi, Wayana, Wayãpi of Camopi), tapereba (Portuguese), tapeliwa (Wayãpi of the upper Oyapock).

U s e s : Cultivated and harvested in the Guianas for its edible, sour fruits, which are sometimes fermented to produce an alcoholic beverage. The leaves, bark and fruits have medicinal applications, locally for the treatment of cancer, dysentery and skin diseases.

3. **Spondias purpurea** L., Sp. Pl., ed. 2: 613. 1762, nom. nov. for *Spondias myrobalanus* L., Fl. Jamaic. 16. 1759. 1725. Lectotype (Bornstein in Howard, Flora Lesser Antilles 5(2): 103. 1989): Sloane, Voy. Jamaica t. 219, fig. 3-5. 1725.   – Plate 9 E-I.

Warmingia pauciflora Engl. in Mart., Fl. Bras. 12(2): 281, t. 57. 1874. Type: Peru, San Martin, Tarapoto, Spruce 4093 (isotypes C, P).

Small tree, 3-15 m, outer bark pinkish-gray to dark gray, smooth or ornamented with spinose, corky outgrowths, inner bark whitish with brown streaks, trichomes whitish, crispate, 0.1-0.2 mm long. Leaves 6-28 cm long, 5-27-foliolate; petiole 2-5.2 cm long, glabrous to sparsely pubescent; rachis 8-20 cm long, glabrous to densely pubescent; leaflets usually subopposite; petiolules pubescent, lateral ones 0-3 mm long, terminal one to 15 mm long; leaflet blades chartaceous, elliptic, ovate to obovate or lanceolate to oblanceolate, usually asymmetrical, 3-6 x 1-2.5 cm, apex obtuse to acute, occasionally retuse, acumen mucronate, base cuneate or attenuate, oblique, margin entire to uncinate-serrulate toward apex, adaxially glabrous except primary vein sparsely to densely pubescent, abaxially glabrous except for basal section of primary vein often sparsely pubescent; primary vein prominulous adaxially, flattened to prominulous abaxially, secondary veins in 5-10 pairs, flattened to prominulous on both surfaces. Inflorescences axillary, racemose-paniculate, 1-10 cm long, few-flowered, developing before leaves are present, glabrous to sparsely pubescent; bracts and bracteoles lanceolate to ovate, 0.5-1.5 mm, apex acute, pubescent abaxially; pedicels 0.5-1.5 mm long, articulate. Flowers with calyx lobes rotund to ovate, 0.8-1.2 mm long, glabrous or very sparsely pubescent abaxially, margin ciliate; petals red to purple, narrowly ovate or oblong, 2.5-3.5 mm long, apically cucullate, glabrous; stamens (8-)10, antepetalous ones 1.3-1.8 mm long, antesepalous ones 1.7-2.2 mm long, anthers oblong, 0.6-0.7 mm long; disk crenulate; pistil 2.3 mm long, ovary subglobose, glabrous, styles 0.5-0.9 mm long, stigmas flattened, capitate. Drupes reddish-orange to purple (typical variety) or yellow (cultivar) at maturity, oblong or obovoid, 2.5-5 x 0.5-3 cm, endocarp oblong, bony.

Distribution: Sporadically cultivated and sometimes adventive in the Old and New World tropics; natural distribution is unclear because of its long association with man. However, it is probably indigenous in tropical dry forests from W Mexico, S to N Peru W of the Andes; cultivated in the Guianas (FG: 7).

Selected specimens: French Guiana: Maripasoula, Fleury 729 (CAY, NY, P, U); Saül in village, ca. 200 m alt., Mori & Mitchell 18777 (NY).

Vernacular names: Guyana: spanish plum (English). French Guiana: mombin rouge (French).

Uses: Cultivated for its edible fruits.

8. **TAPIRIRA** Aubl., Hist. Pl. Guiane 1: 470, 3: t. 188. 1775.
   Type: T. guianensis Aubl.

Small to large trees. Leaves evergreen, alternate, usually imparipinnate, rarely paripinnate; leaflets (sub)opposite, petiolulate, entire; secondary veins usually brochidodromous. Inflorescences terminal or axillary, paniculate, ultimate branches sometimes spicate. Flowers pedicellate or sessile, unisexual, rarely bisexual (plants polygamodioecious), perianth 5-merous, imbricate; corolla greenish yellow, yellow, or cream-colored; disk intra-staminal, 5-10-lobed; stamens (8-)10, in 2 (sometimes strongly) unequal series, filaments subulate; ovary 1-locular, densely pubescent, single ovule apically or subapically suspended, styles (4-)5, stigmas capitate. Drupes oblong-oblique, ellipsoid or globose, mesocarp thin, fleshy, endocarp cartilaginous or bony; embryo curved, cotyledons plano-convex, sometimes with purplish striations.

Distribution: S Mexico to SE Brazil, Bolivia and Paraguay: 7 or more species; in the Guianas 3 species.

Literature: Teichman, I. von. 1990. Pericarp and seed coat structure in Tapirira guianensis (Spondiadeae: Anacardiaceae). S. African J. Bot. 56(4): 435-439.

KEY TO THE SPECIES

1 Leaflets sparsely to densely pubescent with both appressed and erect trichomes, primary vein often densely pubescent adaxially · · 3. *T. obtusa*

Leaflets glabrous or sparsely pubescent, with appressed hairs only, primary veins glabrous adaxially · 2

2 Bark rough, very thick, distinctly laminated, shedding into quadrangular plates; buttresses well developed; rachis horizontally flattened; leaflets glabrous on both surfaces, secondary and tertiary veins usually inconspicuous abaxially; drupes globose · · · · · · · · · · · · *1. T. bethanniana*
Bark usually smooth, thin, not distinctly laminated, not shedding into quadrangular plates; buttresses absent, shallow or rarely well-developed; rachis terete; leaflets glabrous or sparsely pubescent, secondary and tertiary veins prominent abaxially, higher order venation usually conspicuous abaxially; drupes obliquely ovoid to oblong · · · · · · · · · · · *2. T. guianensis*

1. **Tapirira bethanniana** J.D. Mitch., Mem. New York Bot. Gard. 64: 230. 1990. Type: French Guiana, Saül, La Fumée Mt., 200-400 m alt., Mori & Boom 15253 (holotype NY). – Plate 10 A-D, H, I.

Large tree, to 50 m x 80 cm, buttresses well developed, to 1 m tall, outer bark brown, rough, very thick, longitudinally fissured, or shedding into quadrangular plates, inner bark distinctly laminated, pinkish to reddish-brown, trichomes white to golden, mostly appressed, to 0.2 mm long. Leaves 20-41 cm long, 5-11-foliolate (average 7-foliolate); petiole terete, 5.5-8.5 cm long, glabrous; rachis horizontally flattened, glabrous; leaflets usually opposite; petiolules glabrous, lateral ones 1-11 mm long, terminal one 13-30 mm long; leaflet blades usually subcoriaceous to coriaceous, oblong, elliptic, ovate or obovate, 8.5-16.4 x 3.2-7.7 cm, apex acuminate, base attenuate, obtuse, acute or cuneate, frequently oblique, both surfaces glabrous; primary vein impressed or flattened adaxially, prominent abaxially, secondary veins in 7-13 pairs, impressed or flattened adaxially, prominulous abaxially, tertiary and higher order venation usually flattened and inconspicuous ad- and abaxially. Inflorescences subterminal, 10.5-25 cm long, sparsely to densely pubescent; peduncle 0.9-5.5 cm long; bracts lanceolate to deltate, ca. 0.7 mm long, pubescent abaxially; pedicels 1.2-1.4 mm long (staminate flowers), pubescent. Flowers with calyx lobes somewhat fleshy, semi-circular, 1.0 x 0.3 mm, glabrous or sparsely pubescent abaxially, margin somewhat erose; petals oblong, 1.8 x 1.1 mm, glabrous, margin entire or somewhat erose apically; stamens 10, antepetalous filaments 0.6 mm long, antesepalous filaments 0.9 mm long, anthers elliptic, 0.3-0.4 mm long; disk fleshy, annular; pistillode of 4-5 rudimentary styles, 0.5-0.6 mm long, densely pubescent. Pistillate flower not seen. Drupes dark purple to black at maturity, globose, 1.9-2 cm diam.

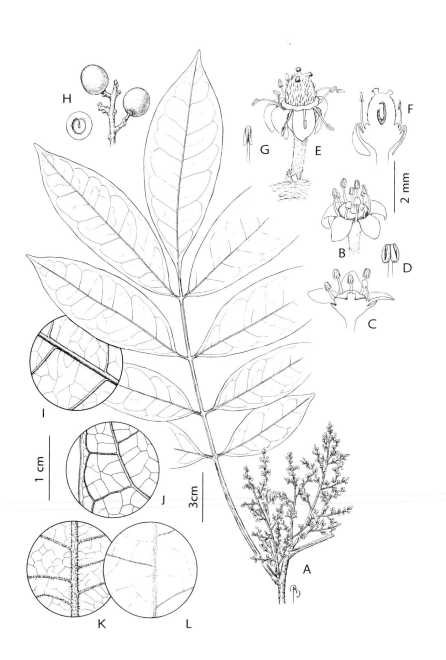

Distribution: Endemic to French Guiana; in primary evergreen, non-inundated high forest, low elevations to 400 m (FG: 18).

Selected specimens: French Guiana: Saül, La Fumée Mt. Trail, ca. 250-400 m alt., Mori & Mitchell 18752 (NY); Piste de St. Elie, Riera & Sabatier 1892 (CAY).

Phenology: Flowering in Nov and fruiting from Dec to Feb.

2. **Tapirira guianensis** Aubl., Hist. Pl. Guiane 1: 470, 3: t. 188. 1775. Type: French Guiana, without locality, Aublet s.n. (holotype BM).
– Plate 10 E-G, J.

Mauria multiflora Mart. ex Benth., Hooker's J. Bot. Kew Gard. Misc. 4: 14. 1852. Type: Martius Herb. Fl. Bras. n. 1274 (holotype K, isotype P).
Tapirira myriantha Triana & Planch., Ann. Sci. Nat. Bot., sér. 5, 14: 295. 1872. Type: Colombia, Valle, Buenaventura, Triana 3693 (holotype COL ?, isotypes BM, K, P, B destroyed).
Tapirira guianensis Aubl. var. *cuneata* Engl. in Mart., Fl. Bras. 12(2): 378. 1876. Type not designated.
Tapirira guianensis Aubl. var. *elliptica* Engl. in Mart., Fl. Bras. 12(2): 378. 1876. – *Rhus amazonica* Poepp., Diar., nom. nud. Type: Poeppig n. 1201 (W).
Tapirira fanshawei Sandwith, Kew Bull. 1955: 470. 1955. Type: Guyana, 90 miles, Bartica-Potaro Road, Fanshawe 2732 (= FD 5531) (holotype K, isotypes K, NY, P, U).

Small to large tree, to 30 m x 60 cm, buttresses absent, shallow or rarely well-developed, outer bark gray or light to dark brown, thin, shallowly fissured or smooth, inner bark pinkish to reddish brown, trichomes whitish to golden, mostly appressed to 0.2 mm long. Leaves 18-70 cm

Plate 10. A comparison of *Tapirira bethanniana* J.D. Mitch. (A-D, H, I) with *Tapirira guianensis* Aubl. (E-G, J) and *Tapirira obtusa* (Benth.) J.D. Mitch. (K-L): A, habit, note the horizontally flattened rachis; B, staminate flower; C, longisection of staminate flower; D, stamen (30x); E, pistillate flower; F, longisection of pistillate flower; G, staminode (30x); H, mature drupe; I, abaxial leaflet surface; J, abaxial leaflet surface; K, abaxial leaflet surface, note the erect, trichomes on primary and secondary veins; L, adaxial leaflet surface, note the densely pubescent primary vein (A-D, I, Mori & Boom 15253, holotype; E-G, Pipoly & Godfrey 7434; H, Sabatier & Riera 2046; J, Mori & Mitchell 18766; K-L, Mori & Mitchell 18775). Reprinted from Mori, S.A. & J.D. Mitchell. 1990. Mem. New York Bot. Gard. 64: 229-234.

long, 5-13-foliolate; petiole 4-20 cm long, glabrous to sparsely pubescent; rachis terete, 4.3-34.5 cm long, glabrous to pubescent; leaflets usually opposite, occasionally subopposite; petiolules glabrous to pubescent, lateral ones 3-15 mm long, terminal one (4-)7-40 mm long; leaflet blades chartaceous to subcoriaceous, narrowly oblong, elliptic, obovate, occasionally narrowly ovate, sometimes falcate, 7.4-24.1 x 2.5-8 cm, apex usually acuminate, base obtuse, acute or cuneate, frequently oblique, adaxially glabrous, abaxially glabrous to sparsely pubescent; primary vein impressed to prominulous adaxially, prominent abaxially, secondary veins in 8-11 pairs, impressed to prominulous adaxially, prominent abaxially, tertiary and higher order venation usually conspicuous abaxially. Inflorescences subterminal, 10.5-31.5 cm long, sparsely to densely pubescent; peduncle ca. 0-8.5 cm long; pedicels 1.0-1.7 mm long (staminate flowers), 1.0-2.5 mm long (pistillate flowers), pubescent. Flowers with calyx lobes somewhat fleshy, semi-circular, 0.4-0.9 x 0.6-0.7 mm, glabrous to sparsely pubescent abaxially, margin somewhat erose; petals oblong, 1.4-2.2 x 0.8-1.0 mm, glabrous, margin entire or somewhat erose apically. Staminate flowers: antepetalous filaments 1.2-1.5 mm long, antesepalous filaments 1.5-1.6 mm long, anthers elliptic, 0.3-0.5 mm long; disk fleshy, annular, 10-crenulate; pistillode of 4-5 rudimentary styles, 0.3-0.6 mm long, densely pubescent. Pistillate flowers with filaments of staminodes 0.8 mm (antepetalous) long and 1.2-1.4 mm (antesepalous) long; ovary barrel-shaped, 0.9-1.2 x 0.7-1.4 mm, densely pubescent, styles 0.1-0.3 mm long, stigmas capitate. Drupes dark purple to black at maturity, oblique-ovoid, ellipsoid or oblong, 1-1.5 x 0.5-1 cm.

Distribution: Costa Rica, S to Bolivia, Paraguay and SE Brazil; abundant in the Guianas; in savannas, semi-deciduous, gallery, disturbed and primary evergreen, non-inundated high forest, alt. 0-600 m (GU: 101; SU: 34; FG: 96).

Selected specimens: Guyana: Between Demerara and Berbice Rs., de la Cruz 1663 (NY); Basin of Rupununi R., Isherton, A.C. Smith 2437 (NY, U). Suriname: Area of Kabalebo Dam project, Nickerie Distr., 30-130 m alt., Lindeman & Görts et al. 118 (NY, U); Tafelberg, Arrowhead Basin, 625 m alt., Maguire 24603 (NY, U). French Guiana: Ile de Cayenne, Anse de Montabo, de Granville 6627 (CAY, NY, P, U); Saül, within 5 km of village, along road to Limonade Cr. from airport, ca. 200-250 m. alt., Mori & Mitchell 18766 (NY).

Phenology: Flowering peaks from Sep to Nov and fruiting peaks from Dec to Feb.

Vernacular names: Guyana: atapirira (Carib.), duka (Arow.), tapiriri (Carib.), warimia (Arow.). Suriname: aganiamaie (Taki-Taki), hoogland witi-oedoe, savanna weti-oedoe, weti-oedoe, witte or zwarte doeka (Sur.). French Guiana: agandjamaí (Boni), aganiamaie (Paramak.), ajawaime (Wayana), axa (Palikur), fatou oudou (Boni), loupe, mombin blanc, mombin faux or fou (Créole), tatapilili (Wayãpi), tatapirica (Brazilian portuguese).

U s e s : Timber of minor importance, more intensively utilized in some other South American countries.

3. **Tapirira obtusa** (Benth.) J.D. Mitch., Novon 3(1): 66. 1993. – *Mauria obtusa* Benth., Hooker's J. Bot. Kew Gard. Misc. 4: 16. 1852. Lectotype (Mitchell, 1993, op. cit.): Guyana, Rob. Schomburgk II 892 (hololectotype K, isolectotype NY). – Plate 10 K-L.

Tapirira pao-pombo Marchand var. *major* Marchand in Warming, Vidensk. Meddel. Dansk Naturhist. Foren. Kjøbenhavn 15: 59. 1873. Type: Brazil, Minas Gerais, Lagoa Santa, Warming s.n. (syntypes C).
Tapirira marchandii Engl. in Mart., Fl. Bras. 12(2): 379. 1876. Based on *Tapirira pao-pombo* Marchand var. *major* Marchand.
Tapirira peckoltiana Engl. in Mart., Fl. Bras. 12(2): 379. 1876. Type: Brazil, Rio de Janeiro, Canta Gallo, Peckolt 348 (holotype BR).

Large tree, to 35 m x 40 cm, buttresses absent or poorly developed, outer bark gray or light brown, thin, longitudinally fissured, inner bark pinkish to reddish-brown, trichomes whitish, golden or brown, appressed and erect to 0.5 mm long. Leaves 16.5-60 cm long, 7-13-foliolate; petiole 4.7-16.5 cm long, sparsely to densely pubescent; rachis 8-35 cm long, usually terete, sparsely to densely pubescent; leaflets (sub)opposite; petiolules usually densely pubescent, lateral ones sessile to 15 mm long, terminal one 20-35 mm long; leaflet blades chartaceous to coriaceous, oblong, elliptic, obovate or ovate, 8-19.7 x 4-6.7 cm, apex acuminate, obtuse or occasionally retuse or emarginate, base usually cuneate, obtuse, acute or rounded, frequently oblique, adaxially glabrous except for densely pubescent midrib and occasionally densely pubescent secondary veins, abaxially sparsely to densely pubescent; primary vein impressed to prominulous adaxially, prominent abaxially, secondary veins in 8-14 pairs, impressed to prominulous adaxially, prominent abaxially, tertiary and higher order venation flattened and usually inconspicuous adaxially, prominulous abaxially. Inflorescences subterminal, 17-27.5 cm long, densely pubescent; peduncle ca. 0-10 mm long; bracts lanceolate to deltate, 0.5-1 mm long, densely pubescent abaxially; pedicels 0.8-1.3 mm long (staminate flowers), or 0.5-(-1.3) mm long (pistillate flowers), densely pubescent. Flowers with calyx lobes somewhat fleshy, semi-circular, 0.3-0.8 x 0.4-0.7 mm, sparsely pubescent abaxially, margin entire; petals oblong, 1.0-1.7 x 0.5-0.6 mm, glabrous adaxially, sparsely pubescent abaxially, margin entire; stamens 10, antepetalous filaments 0.5-1.0 mm long, antesepalous filaments 0.8-1.6 mm long, anthers elliptic, 0.3-0.4 mm long; disk fleshy, annular, 10-crenulate; pistillode of 5 rudimentary styles, 0.3-0.5 mm long, densely pubescent. Pistillate flowers with filaments of staminodes ca. 0.5 mm long; ovary barrel-shaped, 0.9-1.4 x 0.7-1.3 mm, densely pubescent, styles ca. 0.2 mm long, stigmas capitate. Drupes dark purple to black at maturity, globose or ellipsoid, 1.2-2 x 1-1.7 cm.

Distribution: Colombia, Ecuador, Peru, Venezuela, Brazil, the Guianas; less common than *T. guianensis* in the Guianas; in disturbed and primary, evergreen, non-inundated high forest, alt. 0-400 m (GU: 8; SU: 1; FG: 30).

Selected specimens: Guyana: Upper Mazaruni R., village of Kamarang, trail W of airstrip, 505-545 m alt., Boom et al. 8067 (NY); Upper Rupununi R., near Dadanawa, de la Cruz 1402 (F, NY). Suriname: Mapane Cr. area, near Camp 8, LBB 9403 (P, U). French Guiana: Piste de St. Elie, 30 m alt., Bordenave 463 (CAY, NY); Saül, Limonade Trail, 200-400 m alt., Mori, Pipoly & Mitchell 15688 (CAY, NY).

Phenology: Flowering peaks from Sep to Nov and fruiting from Nov to Apr.

Vernacular names: Guyana: duka (Arow.). French Guiana: aganiamaie (Paramak.).

9. **THYRSODIUM** Salzm. ex Benth., Hooker's J. Bot. Kew Gard. Misc. 4: 17. 1852.

Type: T. spruceanum Salzm. ex Benth.

Small to large trees with milky resin. Leaves evergreen, alternate, imparipinnate; leaflets usually alternate, sometimes subopposite, entire, petiolulate. Inflorescences terminal and/or axillary, thyrsoid; bracts and bracteoles deciduous. Flowers pedicellate, unisexual (plant dioecious), perigynous, hypanthium deeper in staminate than in pistillate flowers; perianth 5-merous; calyx lobes valvate; petals imbricate; disk adnate to hypanthium; stamens 5, alternating with petals, filaments subulate; staminodes either resembling reduced stamens or modified into short lobes on rim of hypanthium, filaments short or obsolete, anthers sometimes pubescent; pistillode rudimentary in staminate flowers; ovary globose to ovoid, 1-locular, with 1 lateral (sub-basal to sub-apical) ovule, style and rudimentary style terminal, simple or 2-3-

branched, stigmas 1-3. Drupes ovoid, ellipsoid, oblong, or globose, apex somewhat cuspidate, mesocarp fleshy, endocarp crustaceous; embryo straight, cotyledons plano-convex.

Distribution: Amazonian Colombia, Peru, Bolivia, S and E Venezuela, the Guianas, Amazonian and E Brazil: 6-7 species; in the Guianas 3 species.

Literature: Mitchell, J.D. & D.C. Daly. 1993. A revision of Thyrsodium (Anacardiaceae). Brittonia 45(2): 115-129.

KEY TO THE SPECIES

1 Leaflets sparsely to densely pubescent; petals 2.6-3.8 mm long; pistillode, pistil and fruits pubescent · *3. T. spruceanum*
Leaflets (sub)glabrous; petals 2-2.5 mm long; pistillode, pistil and fruits glabrous · 2

2 Leaflet apex usually short-acuminate, rounded or emarginate, tertiary venation prominulous adaxially; staminodes 0.3-0.4 mm long; stigmas 2-3 · *1. T. guianense*
Leaflet apex usually long-acuminate, tertiary venation flat adaxially; staminodes 0.1-0.2 mm long; stigma 1 · · · · · · · · · · · · · *2. T. puberulum*

1. **Thyrsodium guianense** Sagot ex Marchand, Rev. Anacardiac. 160. 1869. – *Garuga guianensis* (Sagot) Engl. in Mart., Fl. Bras. 12(2): 288. 1874. Lectotype (Mitchell & Daly, Brittonia 45(2): 128. 1993): French Guiana, Acarouany (Karouany), Sagot 1202 (hololectoype P, isolectotypes BM, K, U). – Plate 11 A-C.

Medium-sized tree, to 36 m x 50 cm, with low buttresses, outer bark smooth, gray, thin, inner bark reddish-orange, trichomes golden, to 0.25 mm long. Leaves 19.5-47.5 cm long, 6-14-foliolate; petiole 5-13 cm long, pubescent; rachis 5.5-24 cm long, pubescent; leaflets opposite to alternate; petiolules glabrous to sparsely pubescent, lateral ones 5-10 mm long, terminal one 8-30 mm long; leaflet blades coriaceous, often shining adaxially, broadly to narrowly obovate, ovate, elliptic or rarely oblong, 7.3-12.2 x 3.1-5.3 cm, apex short-acuminate, rounded-truncate, emarginate or retuse, very rarely long-acuminate, base usually cuneate, sometimes acute or obtuse, margin frequently revolute, both surfaces glabrous or abaxially sparsely pubescent; primary vein impressed adaxially, prominent abaxially, secondary veins in 8-11 pairs, flat to prominulous adaxially, prominulous to prominent abaxially, tertiary

42

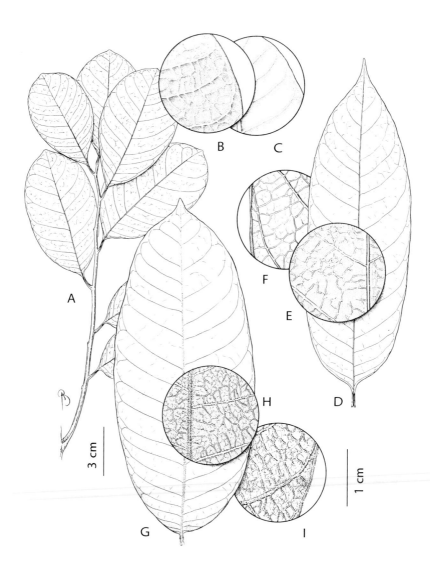

Plate 11. A comparison of *Thyrsodium* leaves, *T. guianense* Sagot ex Marchand (A-C), *T. puberulum* J.D. Mitch. & Daly (D-F), *T. spruceanum* Salzm. ex Benth. (G-I): A, habit; B, adaxial leaflet surface; C, abaxial leaflet surface; D, leaflet; E, adaxial leaflet surface; F, abaxial leaflet surface; G, leaflet; H, adaxial leaflet surface, note the frequently densely pubescent primary vein; I, abaxial leaflet surface (A-C, de Granville 3392; D-F, Boom 2006; G-I, Lindeman 4773).

veins prominulous adaxially, prominulous but less conspicuous abaxially. Inflorescences densely pubescent, 12-35 cm long; peduncle 1.5-4.7 cm long; pedicels 1-3 mm long, densely pubescent. Flowers with hypanthium cupular; calyx lobes deltate or ovate, 1.1-2 x 1.3-1.6 mm, densely pubescent abaxially; petals white when fresh, ovate to narrowly ovate, 2.2-2.5 x 1.3 mm, densely pubescent adaxially, sparsely pubescent abaxially. Staminate flowers: filaments 0.5-0.8 mm long, sparsely pubescent, anthers oblong, 0.8-1 mm long; pistillode cylindrical, 2-2.5 mm long, glabrous, crowned by 2-3 stigmatic lobes. Pistillate flowers: staminodes 0.3-0.4 mm long, rudimentary anthers ovoid or oblong; ovary ovoid, 1.8 x 1.5-2.1 mm, glabrous, style ca. 1 mm long, crowned by 2-3 capitate stigmas. Drupes bluish-green (at maturity), obovoid, ellipsoid or oblong, 2.5 x 1.4 cm, glabrous.

Distribution: Suriname, French Guiana, Brazil (Amapá and adjacent Pará); disturbed and primary evergreen, non-inundated high forest, alt. 0-400 m (SU: 7; FG: 17).

Selected specimens: Suriname: Brokopondo Distr., 8 km ESE of Brownsweg, van Donselaar 1905 (U); Moengo Tapoe-Grote Zwiebelzwamp road, Lanjouw & Lindeman 403 (K, U). French Guiana: Ile de Cayenne, Mt. Mahury, S of Lac Rorota, de Granville BC.55 (NY, P, U); Saül, near Eaux Claires on Sentier Botanique, Mori et al. 21687 (NY).

Phenology: Flowering from Oct to Dec and fruiting from Nov to Dec.

Vernacular names: French Guiana: encens rouge (French), grand moni (Paramak.), kanamboeli tingimoni (Paramak.?).

2. **Thyrsodium puberulum** J.D. Mitch. & Daly, Brittonia 45(2): 122. 1993. Type: French Guiana, Saül, Tracé Limonade, de Granville B.5486 (holotype NY, isotype CAY). – Plate 11 D-F, 12.

Small to large tree, 8-35 m x 50 cm, with shallow buttresses to 50 cm high, outer bark brown, thin, longitudinally fissured, inner bark reddish-orange, trichomes brown, appressed, to 0.3 mm long. Leaves 29-93 cm long, 8-11-foliolate; petiole 8.7-22 cm, glabrous to sparsely pubescent; leaflets alternate or subopposite; petiolules glabrous to sparsely pubescent, lateral ones 6-10 mm long, terminal one 9-25 mm long; leaflet blades usually coriaceous, often shiny adaxially, narrowly oblong, lanceolate or occasionally narrowly elliptic, ovate or obovate, 10.2-27.5 x 4.2-10.2 cm, apex long-acuminate, base rounded to obtuse or occasionally cuneate, sometimes oblique, margin sometimes revolute, adaxially glabrous, abaxially glabrous to sparsely pubescent;

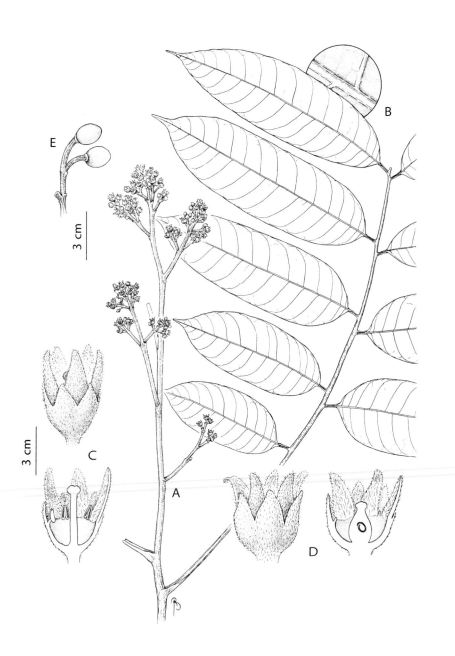

primary vein impressed to prominulous adaxially, prominent abaxially, secondary veins in 12-17 pairs, flat to prominulous adaxially, prominent abaxially, tertiary venation flat and discolorous adaxially, prominulous abaxially. Inflorescences 18-35 cm long, sparsely to densely pubescent; peduncle 3.3-3.5 cm long; pedicels 2.5-3.5 mm long. Flowers with hypanthium cupular; calyx lobes ovate or deltate, 2-2.3 x 1.5-2.2 mm, densely pubescent abaxially; petals greenish white to pale yellow when fresh, narrowly ovate, 3.2-3.6 x 1.5-1.7 mm, densely pubescent on both surfaces. Staminate flowers: corolla greatly exceeding calyx; filaments 0.2-0.25 mm long, glabrous; anthers oblong, 1-1.3 mm long, glabrous or with a few trichomes; pistillode cylindrical, 3-3.5 mm long, glabrous, rudimentary stigma capitate, sometimes apically notched. Pistillate flowers: corolla not greatly exceeding calyx; staminodes present, 0.1-0.2 mm long, without or provided with rudimentary anthers; ovary ovoid, 2.3 x 2.4 mm, glabrous, style 0.5 mm long, stigma 1, capitate. Drupes color when fresh unknown, ellipsoid to subglobose, 2-3 cm diam., glabrous.

Distribution: NE Suriname, French Guiana, Brazil (Amapá and Pará); disturbed and primary, evergreen, non-inundated, high-forest, alt. 0-400 m (SU: 1; FG: 25).

Selected specimens: Suriname: B.S.H. ekspl. Patamacca, footpath, km 8.1, Borsboom in LBB 12005 (U). French Guiana: Saül, La Fumée Mt., Boom et al. 2006 (NY); Right bank of Camopi R., tributary of Oyapock R., at Roche José, 800 m from shore, Oldeman 2582 (NY, P, U).

Phenology: Flowering from Sep to Oct and fruiting from Oct to Jan.

Vernacular names: French Guiana: carapa oyac (language?), encens rouge (French), grand moni, moni (Paramak.), taite tukasina (Wayãpi), tingimoni (Créole).

Plate 12. *Thyrsodium puberulum* J.D. Mitch. & Daly: A, habit; B, detail of abaxial leaflet surface; C, external view and longisection of staminate flower, showing corolla greatly exceeding calyx and subsessile anthers; D, external view and longisection of pistillate flower, showing corolla not greatly exceeding calyx, staminodes without rudimentary anthers, and glabrous ovary; E, fruits (A-B & D, Prance & N.T. Silva 58888; C, Irwin et al. 48121; E, Rabelo et al. 3113). Reprinted from Mitchell, J.D. & D.C. Daly. 1993. Brittonia 45(2): 115-129, fig. 4.

3. **Thyrsodium spruceanum** Salzm. ex Benth., Hooker's J. Bot. Kew Gard. Misc. 4: 17. 1852. – *Garuga spruceana* (Benth.) Engl. in Mart., Fl. Bras. 12(2): 286. 1874. Type: Brazil, Pará, in campos near Santarém, Spruce 793 (holotype K, isotypes GH, NY).

– Plate 11 G-I.

Thyrsodium schomburgkianum Benth., Hooker's J. Bot. Kew Gard. Misc. 4: 18. 1852. – *Garuga schomburgkiana* (Benth.) Engl. in Mart., Fl. Bras, 12(2): 287. 1874. Type: Guyana, without locality, Rob. Schomburgk I 892 (holotype K, isotypes BM, NY).
Thyrsodium paraense Huber, Bull. Soc. Bot. Genève, sér. 2, 6: 183. 1914. Type: Brazil, Pará, forests adjacent to R. Cuminá-mirim, Ducke MG 7969 (holotype MG).
Thyrsodium dasytrichum Sandwith, Bull. Misc. Inform. 1932: 210. 1932. Type: Guyana, Cuyuni R., near Upper Camaria, Lockie s.n. (= FD 2019) (holotype K, isotype NY).

Small to medium-sized tree, 3-25 m tall, to 100 cm diam., shallowly buttressed or unbuttressed, outer bark gray-brown, thin, hard, very shallowly fissured, inner bark dark reddish-orange, trichomes ferrugineous to purple brown, to 0.5 mm long. Leaves 24-83.5 cm long, 7-15-foliolate; petiole densely pubescent, 4-17.5 cm long; rachis sparsely to densely pubescent, 10-52.8 cm long; petiolules densely pubescent, lateral ones 3-15 mm long, terminal one 18-38 mm long; leaflet blades chartaceous to subcoriaceous, shiny or dull adaxially, narrowly oblong, lanceolate or occasionally narrowly elliptic, ovate, obovate or oblanceolate, 9-25.5 x 3.3-9.3 cm, apex short- to long-acuminate, base obtuse, rounded, truncate, cordate, sometimes oblique, margin sometimes revolute, also sometimes sparsely ciliate, both surfaces sparsely to densely pubescent; primary vein impressed adaxially, prominent abaxially, secondary veins in 9-19 pairs, often impressed to prominulous adaxially, prominent abaxially. Inflorescences 12.5-64 cm long, sparsely to densely pubescent; peduncle 1.5-8 cm long; pedicels 2-4 mm long. Flowers with calyx lobes ovate or deltate, 2.1-3.5 x 0.9-2.2 mm, densely pubescent abaxially; petals cream-colored to pale yellow when fresh, narrowly ovate, 2.6-3.8 x 1.3-2 mm, sparsely to densely pubescent adaxially, pubescent abaxially. Staminate flowers: hypanthium cupular; filaments 0.8-1.3 mm long, sparsely pubescent, anthers oblong, 0.9-1.3 mm long, glabrous; pistillode cylindrical, 2.5-4 mm long, sparsely to densely pubescent, rudimentary stigmas 2-3-lobed or subcapitate. Pistillate flowers: hypanthium more shallowly cupular; rudimentary anthers sagittate, 1 mm long; ovary globose, 2.4-5 x 2.3-4.2 mm, densely pubescent, style 1 mm long, crowned by a capitate or 2-lobed stigma. Drupes green to bluish-green when ripe, ovoid, ellipsoid, or oblongoid, 1.7-2.5 x 2-3 cm, densely pubescent.

Distribution: Venezuela (S of Orinoco R.), Amazonian Brazil, the Guianas, disjunct to the Serranía de la Macarena of Colombia and the Atlantic forest of Brazil (NE Brazil S to Espírito Santo); disturbed and primary, evergreen, non-inundated, high forest, alt. 0-600 m (GU: 6; SU: 6; FG: 3).

Selected specimens: Guyana: Upper Mazaruni R., Kamarang, trail W of airstrip, Boom et al. 8195 (NY); NW portion of Kanuku Mts., Mt. Iramaikpang, A.C. Smith 3609 (NY, U). Suriname: Jodensavanne-Mapane Cr. area, Suriname R., Lindeman 4773 (NY, U); Nickerie Distr., area of Kabalebo Dam project, Lindeman & de Roon 7143 (U). French Guiana: Saül, La Fumée Mt., Boom & Mori 2439 (CAY, NY); Nouragues Station, Basin of Arataye R., Sabatier & Prévost 2838 (CAY, NY).

Phenology: Flowering from Sep to Apr.

WOOD AND TIMBER

by

BEN J.H.TER WELLE[2], PIERRE DÉTIENNE[3] AND T. TERRAZAS[4]

WOOD ANATOMY

FAMILY CHARACTERISTICS

Vessels diffuse, solitary and in irregular clusters and radial multiples of 2-8, perforations simple, intervascular pits alternate, round to polygonal, sometimes elongated, 7-13(-16) μm.

Rays almost exclusively uniseriate in most samples of *Anacardium*, but commonly few uniseriate and 2-4(-5)-seriate in other genera, heterogeneous. Rhombic crystals in most genera, silica grains in *Anacardium*, and *Loxopterygium*. Radial intercellular canals common in most genera, but not observed in *Anacardium*, and *Mangifera*.

Parenchyma paratracheal scanty vasicentric, and sometimes lozenge-aliform in *Anacardium*, and *Mangifera*, occasionally few rhombic crystals.

Fibres non-septate, septate, or a mixture of both, with small simple or minutely bordered pits.

Special characteristics

Radial intercellular canals

Presence or absence of radial canals in rays is a constant character at generic level (Hess, 1946; Dong & Baas, 1993; Terrazas, 1994; Terrazas & Dickison, in press).

Radial intercellular canals are absent in *Anacardium*, and *Mangifera*, and present in other genera studied. Size, anatomy and distribution of these radial canals is variable. For details see Table 1. See also Figs. 8, 13, 14, and 19.

[2] Herbarium Division, Department of Plant Ecology and Evolutionary Biology, Heidelberglaan 2, 3584 CS Utrecht, The Netherlands.

[3] C.I.R.A.D.-Forêt, Maison de la Technologie, BP 5035, Montpellier, Cedex 1, France.

[4] Collegio de Postgraduados, Institucion de Ensenanza e Investigacion en Ciencias. Agricolas, Centro de Botánica, Montecillo, Edo. de Mexico, 56230, Mexico.

Table 1. Radial intercellular canals in Guianan Anacardiaceae

Genus	Diam. (µm) (Tang.sect.)	Shape	Distribution	Anatomy
Astronium	20-40	round	common	2-3 layers of lignified sheath cells, to oval and occasionally 1 layer of unlignified very thin-walled epithelial cells
Cyrtocarpa	25-35	round	common	1(-2) layers of lignified epithelial cells
Loxopterygium	20x60/ 75x165	oval to slightly oval	common	1 layer of unlignified very thin-walled epithelial cells, surrounded by 1(-2) layers of lignified sheath cells
Spondias	40x50/ 55-85	slightly oval or round	common	1-2 layers of lignified epithelial cells
Tapirira	35x55/ 90x140	oval, sometimes slightly oval	common	1 layer of unlignified very thin-walled epithelial cells sometimes present, surrounded by 1-(4) layers of lignified sheath cells
Thyrsodium	30x65/ 90-160	oval and sometimes	common to scarce	1 layer of unlignified very thin-walled epithelial cells sometimes present, surrounded by 2-4 layers of lignified sheath cells

Fibres / Parenchyma

In some genera, such as *Astronium, Cyrtocarpa, Loxopterygium,* and *Tapirira* all fibres are septate. In *Anacardium, Spondias,* and *Thyrsodium* septate and non-septate fibres occur. Finally, in *Mangifera* all fibres are non-septate. Septate and non-septate fibres are sometimes randomly distributed. However, in e.g. *Thyrsodium* septate fibres are only found near vessels, and in combination with parenchyma. See also Figs. 6, 7, 8, 9, 12, 13, and 14.

KEY FOR IDENTIFICATION OF GUIANAN GENERA

1 Silica grains present in ray cells $\cdots\cdots\cdots\cdots\cdots\cdots\cdots\cdots\cdots$ 2
  Silica grains absent $\cdots\cdots\cdots\cdots\cdots\cdots\cdots\cdots\cdots\cdots\cdots\cdots$ 3

2 Intercellular canals present in the rays; over 10 vessels per sq. mm;
  fibres septate $\cdots\cdots\cdots\cdots\cdots\cdots\cdots\cdots\cdots$ *Loxopterygium sagotii*
  Intercellular canals absent; less than 5 vessels per sq. mm; fibres
  generally non-septate $\cdots\cdots\cdots\cdots\cdots\cdots\cdots\cdots\cdots$ *Anacardium*

3 Intercellular canals present in the rays $\cdots\cdots\cdots\cdots\cdots\cdots\cdots\cdots$ 4
  Intercellular canals absent $\cdots\cdots\cdots\cdots\cdots\cdots\cdots\cdots$ *Mangifera indica*

4 Fibres all septate $\cdots\cdots\cdots\cdots\cdots\cdots\cdots\cdots\cdots\cdots\cdots\cdots$ 5
  Fibres only in part septate $\cdots\cdots\cdots\cdots\cdots\cdots\cdots\cdots\cdots\cdots$ 7

5 Number of vessels over 20 per sq. mm; 2-4 septa per fibre $\cdots\cdots\cdots$
  $\cdots\cdots\cdots\cdots\cdots\cdots\cdots\cdots\cdots\cdots\cdots$ *Cyrtocarpa velutinifolia*
  Number of vessels less than 16 per sq. mm; 1-3 septa per fibre $\cdots$ 6

6 Intervascular pits 10-13 µm; rays over 600 µm in height $\cdots$ *Tapirira*
  Intervascular pits 7-11 µm; rays under 600 µm in height $\cdots$ *Astronium*

7 Number of rays 2-3(0-4) per mm; septate fibres distribution at
  random; rhombic crystals common in ray cells $\cdots$ *Spondias mombin*
  Number of rays 6-7(3-10) per mm; septate fibres distribution
  restricted to parenchyma/vessel area; rhombic crystals absent or rare
  in ray cells $\cdots\cdots\cdots\cdots\cdots\cdots\cdots\cdots\cdots\cdots\cdots\cdots$ *Thyrsodium*

GENERIC DESCRIPTIONS

ANACARDIUM L. – Figs. 1, 2, 3.

Vessels diffuse, solitary (35-50%), and in irregular clusters of 2-3, round
to oval, 1-4(0-5) per sq. mm, diameter variable, in two size groups,
respectively 150-195(120-210) µm and 80-105(70-115) µm. Vessel-
member length: 398-621(253-805) µm. Perforations simple.
Intervascular pits alternate, oval or polygonal, 10-13 µm. Vessel-ray and
vessel-parenchyma pits large, simple, round, elongated, gash-like to
occasionally scalariform. Thin-walled tyloses sometimes present.
Rays almost exclusively uniseriate, few with a small biseriate part, or the
majority biseriate in one sample of *A. giganteum*, 7-8(5-11) per mm, up
to 800-1150 µm (= 22-26 cells) high. Heterogeneous, composed of
variable amounts of square, upright, slightly procumbent and procumbent
cells. Silica grains round, 3-8 mm in diameter, present in ray cells.

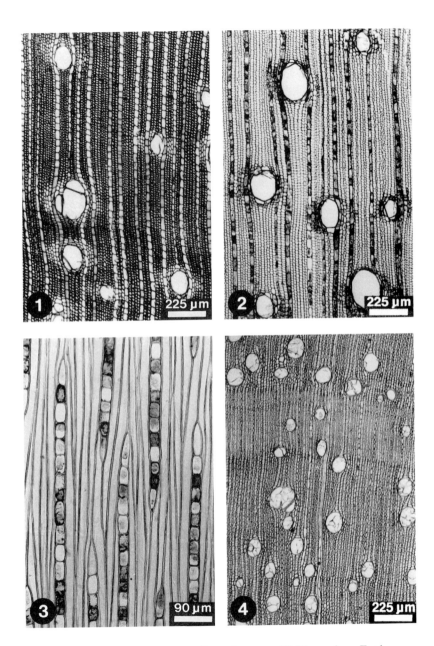

Fig. 1. Transverse section. *Anacardium giganteum* W. Hancock ex Engl.
Fig. 2. Transverse section. *Anacardium giganteum* W. Hancock ex Engl.
Fig. 3. Tangential section. *Anacardium giganteum* W. Hancock ex Engl.
Fig. 4. Transverse section. *Astronium lecointei* Ducke

Parenchyma paratracheal, lozenge-aliform to occasionally confluent. Strands of 2-5 cells.

Fibres generally non-septate, but some septate fibres observed in *A. spruceanum* and *A. giganteum*, lumen up to 15-21 µm, walls up to 4 µm. Pits simple or minutely bordered, on radial walls and few on tangential walls, 4-5 µm. Gelatinous fibres abundant. Length: 879-1170(690-1460) µm. F/V ratio: 1.75-2.21.

Studied: *A. giganteum, A. occidentale, A. spruceanum.*

ASTRONIUM Jacq. – Figs. 4, 5, 6, 7, 8, 9.

Growth rings present, sometimes absent, indicated by differences in lumen diameter of fibres and in some samples also by a slight difference in vessel diameter.

Vessels diffuse, solitary (40-70%) and in radial multiples and irregular clusters of 2-6(-8), round to slightly oval, 4-14(2-21) per sq. mm, diameter 100-160(90-175) µm. Vessel-member length: 506-525(385-620) µm. Perforations simple. Intervascular pits alternate, oval to polygonal, occasionally round, 7-11 µm. Vessel-ray and vessel-parenchyma pits large, simple or with very reduced borders, elongated, sometimes nearly scalariform. Thin-walled tyloses common.

Rays uniseriate, and 2-4-seriate with uniseriate margins of 1-8 cells, 4-8(2-9) per mm, up to 400-575 µm (= 18-38 cells) high. Heterogeneous, composed of procumbent cells, and square and upright cells in the margins. Rhombic crystals common to abundant in slightly enlarged marginal ray cells, but absent or very few in procumbent cells. Radial intercellular canals common.

Parenchyma scarce, paratracheal, scanty vasicentric. Strands of 2-4 cells. In one sample a few rhombic crystals.

Fibres septate, 1-3 septa per fibre, lumen up to 7-20 µm, walls up to 2-4 µm. Pits simple, on radial walls and few on tangential walls, 2-3 µm. Length: 1005-1177(775-1360) µm. F/V ratio: 1.91-2.33.

Studied: *A. lecointei, A. ulei.*

CYRTOCARPA Kunth – Fig. 10.

Vessels diffuse, solitary (50-55%), and in irregular clusters and radial multiples of 2-4(-8), round or slightly angular, 23(17-29) per sq. mm, diameter 115(90-160) µm. Vessel-member length: 560(353-764) µm. Perforations simple. Intervascular pits alternate, polygonal and occasionally elongated, 10-11 and 16x10 µm, respectively. Vessel-ray

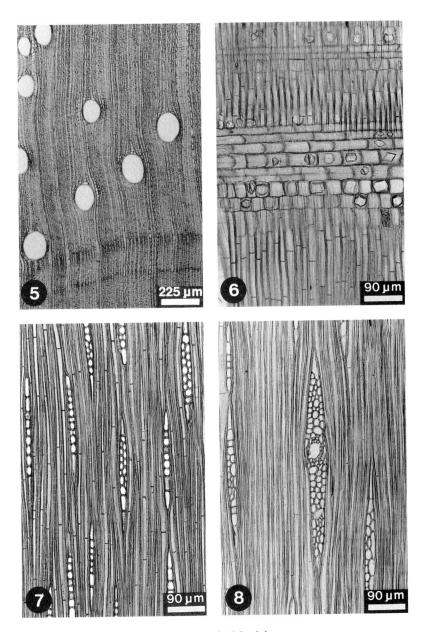

Fig. 5. Transverse section. *Astronium ulei* Mattick
Fig. 6. Radial section. *Astronium lecointei* Ducke. Rhombic crystals in the ray cells.
Fig. 7. Tangential section. *Astronium ulei* Mattick. Septate fibres.
Fig. 8. Tangential section. *Astronium ulei* Mattick. Radial intercellular canal.

and vessel-parenchyma pits large, simple, round but mostly elongated, gash-like to occasionally scalariform.

Rays 2-3-seriate, with uniseriate margins of 1-2 cells, 4 per mm, up to 430 µm (= 15 cells) high. Heterogeneous, composed of procumbent cells, and square and upright cells in uniseriate margins. Rhombic crystals common in square and upright ray cells. Radial intercellular canals common.

Parenchyma paratracheal, scanty vasicentric. Strands of 3-5 cells.

Fibres septate, 2-4 septa per fibre, lumen up to 15 µm, walls up to 3 µm. Pits simple, on radial walls and few on tangential walls, 2 µm. Gelatinous fibres abundant. Length: 1067(882-1558) µm. F/V ratio: 1.91.

Studied: *C. velutinifolia.*

LOXOPTERYGIUM Hook. f. – Figs. 11, 12, 13, 14, 15.

Growth rings absent or occasionally present, due to differences in vessel diameter.

Vessels diffuse, solitary (25-35%), and in radial multiples and irregular clusters of 2-8(-10), round and slightly oval, 11-18(5-22) per sq. mm, diameter 120-155(95-195) µm. Vessel-member length: 441-550(294-647) µm. Perforations simple. Intervascular pits alternate, round or polygonal, 7-11 µm. Vessel-ray and vessel-parenchyma pits large, simple, elongated, gash-like to scalariform. Thin-walled tyloses sometimes present.

Rays uniseriate and 2(-3)-seriate, with uniseriate margins of 1-5 cells, 5-7(4-8) per mm, up to 350-460 µm (= 14-20 cells) high. Heterogeneous, composed of procumbent cells, and square and upright cells at the margins. Radial intercellular canals common. Silica grains round, diameter 5-8 µm, present mainly in procumbent ray cells.

Parenchyma scarce, paratracheal, scanty vasicentric. Strands of 3-5 cells.

Fibres septate, 1-3 septa per fibre, lumen up to 16-21 µm, walls up to 2-5 µm. Pits simple, on radial walls and few on tangential walls, 2-3 µm. Gelatinous fibres abundant. Length: 970-1200(790-1495) µm. F/V ratio: 1.98-2.38.

Studied: *L. sagotii.*

MANGIFERA L. – Figs. 16, 17.

Growth rings sometimes present, as a result of differences in vessel diameter, in combination with prominent tangential parenchyma bands.

Vessels mostly diffuse, solitary (30%) and in irregular clusters and radial multiples of 2-6, round to oval, 4(2-6) per sq. mm, diameter 125-155(70-190) µm. Vessel-member length: 300-490(205-625) µm.

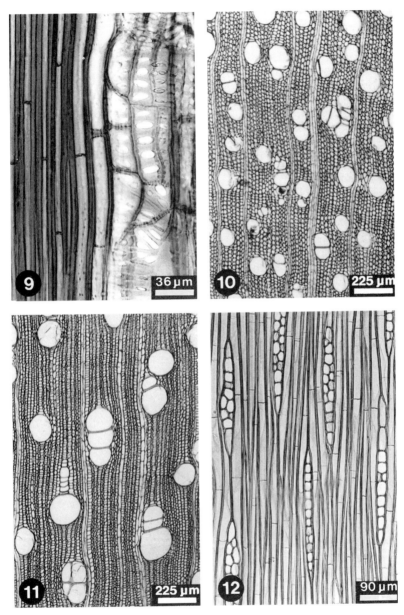

Fig. 9. Radial section. *Astronium ulei* Mattick. Septate fibres and vessel-
     parenchyma pits.
Fig. 10. Transverse section. *Cyrtocarpa velutinifolia* (R.S. Cowan) J.D. Mitch.
     & Daly.
Fig. 11. Transverse section. *Loxopterygium sagotii* Hook. f.
Fig. 12. Tangential section. *Loxopterygium sagotii* Hook. f. Septate fibres.

56

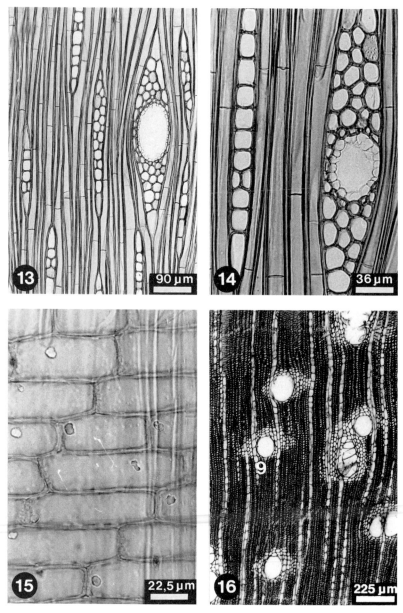

Fig. 13. Tangential section. *Loxopterygium sagotii* Hook. f. Radial intercellular canal and septate fibres.

Fig. 14. Tangential section. *Loxopterygium sagotii* Hook. f. Radial intercellular canal with distinctive unlignified epithelial cells.

Fig. 15. Radial section. *Loxopterygium sagotii* Hook. f. Small silica grains in the ray cells.

Fig. 16. Transverse section. *Mangifera indica* L.

Perforations simple. Intervascular pits alternate, polygonal, sometimes round, 9-14 µm. Vessel-ray and vessel-parenchyma pits large, simple, round, elongated to occasionally scalariform. Thin-walled tyloses sometimes present.

Rays uniseriate and biseriate, 8(7-9) per mm, up to 300 µm (= 7 cells) high. Heterogeneous, composed of square, upright and procumbent cells. Rhombic crystals common in all cell types.

Parenchyma paratracheal, lozenge- and winged-aliform, sometimes apotracheal tangential bands, variable in width. Strands of 2-3 cells.

Fibres non-septate, lumen up to 14 µm, walls up to 2-3 µm. Pits simple, on radial and tangential walls, 2-3 µm. Gelatinous fibres abundant. Length: 490-790(470-1245) µm. F/V ratio: 1.63.

S t u d i e d : *M. indica* (cultivated in Rupununi Savanna).

SPONDIAS L. – Figs. 18, 19.

Vessels diffuse or occasionally slightly semi-ringporous, solitary (40-50%) and in irregular clusters and radial multiples of 2-4(-8), round to slightly oval, 4-8(2-12) per sq. mm, diameter 185-225(140-290) µm, occasionally a few much smaller. Vessel-member length: 529-685(412-823) µm. Perforations simple. Intervascular pits alternate, polygonal, 10-13 µm. Vessel-ray and vessel-parenchyma pits large, simple, round, elongated, and gash-like. Thin-walled tyloses sometimes present.

Rays 3-5(-6)-seriate, with uniseriate margins of 1-2 cells, and very few uniseriates, 2-3(0-4) per mm, up to 690-1150 µm (= 27-60 cells) high. Heterogeneous, composed of procumbent cells, and square and upright cells at the margins. Rhombic crystals common, restricted to marginal square and upright cells. Radial intercellular canals common.

Parenchyma paratracheal, scanty to vasicentric. Strands of 3-4 cells. Rhombic crystals sometimes present.

Fibres septate (1-3 septa) and non-septate, in variable amounts, lumen up to 24 µm, walls up to 4 µm. Pits simple or minutely bordered, on radial walls and few on tangential walls, 1-2.5 µm. Gelatinous fibres abundant. Length: 1313-1689(1147-2234) µm. F/V ratio: 2.42-2.65.

S t u d i e d : *S. mombin.*

TAPIRIRA Aubl. – Figs. 20, 21, 22.

Growth rings present, but often faint, due to differences in fibre lumen diameter, diameter of vessels, or a combination of these two characteristics.

58

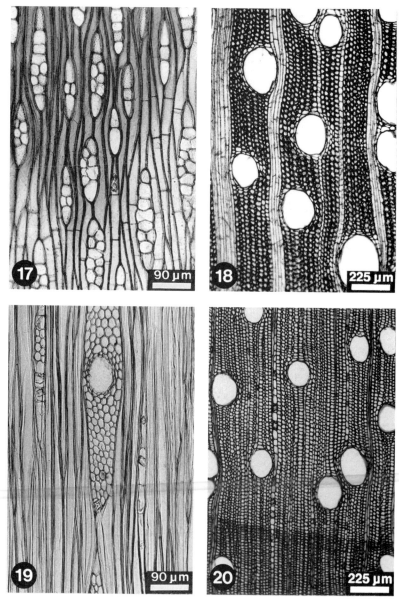

Fig. 17. Tangential section. *Mangifera indica* L.
Fig. 18. Transverse section. *Spondias mombin* L.
Fig. 19. Tangential section. *Spondias mombin* L. Rhombic crystals in the ray
cells and intercellular canal.
Fig. 20. Transverse section. *Tapirira bethanniana* J.D. Mitch.

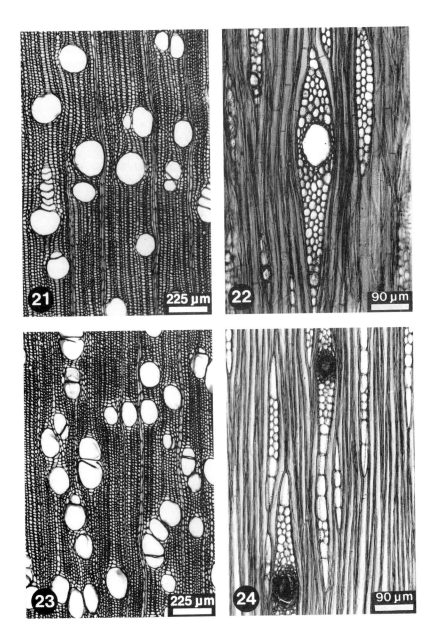

Fig. 21. Transverse section. *Tapirira bethanniana* J.D. Mitch.
Fig. 22. Tangential section. *Tapirira guianensis* Aubl. Radial intercellular canal.
Fig. 23. Transverse section. *Thyrsodium puberulum* J.D. Mitch. & Daly.
Fig. 24. Tangential section. *Thyrsodium spruceanum* Salzm. ex Benth.

Vessels diffuse, solitary (30-80%), and in radial multiples and irregular clusters of 2-5(-8), round to slightly oval, 7-13(5-21) per sq. mm, diameter 140-170(115-200) μm. Vessel-member length: 556-635(353-882) μm. Perforations simple. Intervascular pits alternate, polygonal, 10-13 μm. Vessel-ray and vessel-parenchyma pits large, simple, elongated, and gash-like, occasionally almost scalariform. Thin-walled tyloses sometimes present.

Rays 2-4-seriate, with uniseriate margins of 1-5(-7) cells, and few uniseriate, 5-6 per mm, up to 620-800 μm (= 25-35 cells) high. Heterogeneous, composed of procumbent cells, and square and upright cells at the margins, 1-5(-7) cells high. Rhombic crystals common, restricted to square and upright cells. Radial intercellular canals common, when present these rays are (much) wider locally than other multiseriate rays.

Parenchyma paratracheal, vasicentric. Strands of 3-4(-5) cells.

Fibres all septate, 1-3 per fibre, lumen up to 15-21 μm, walls up to 2.5-4 μm. Pits simple, on radial walls and few on tangential walls, small, less than 2 μm. Length: 1189-1348(940-1617) μm. F/V ratio: 1.87-2.27.

S t u d i e d : *T. bethanniana, T. guianensis.*

THYRSODIUM Salzm. ex Benth. – Figs. 23, 24.

Growth rings absent or very faint.

Vessels diffuse, solitary (50-75%), and in radial multiples and irregular clusters of 2-5(-8), round, 13-17(8-22) per sq. mm, diameter 100-145(90-160) μm. Vessel-member length: 551-601(441-764) μm. Perforations simple, few scalariform or foraminate. Intervascular pits alternate, round/oval and occasionally polygonal, 8-11 μm. Vessel-ray and vessel-parenchyma pits large, simple or with very much reduced borders, round, elongated, and gash-like.

Rays 2-3-seriate, with uniseriate margins of 1-5(-10) cells, 6-7(3-10) per mm, up to 560-900 μm (= 17-30 cells) high. Heterogeneous, composed of procumbent cells, and square and upright cells at the margins. Radial intercellular canals present, in some samples scarce. Rhombic crystals absent or present (but few) in upright or square cells.

Parenchyma paratracheal, scanty to vasicentric. Strands of 2-4 cells.

Fibres non-septate and septate when situated near vessel and parenchyma, lumen up to 11-16 μm, walls up to 2-3 μm. Pits simple or minutely bordered, on radial walls and few on tangential walls, small, 2 μm. Gelatinous fibres rare. Length: 1036-1179(853-1352) μm. F/V ratio: 1.93-1.96.

S t u d i e d : *T. guianense, T. puberulum, T. spruceanum.*

TIMBERS AND THEIR PROPERTIES

Mangifera indica and *Schinus terebinthifolia* are species introduced in the Guianas, for their fruits. Consequently, these species are of no commercial value for the timber industry. *Cyrtocarpa velutinifolia* is a rare, small tree with no commercial value either.
Some species of *Anacardium*, *Astronium*, *Loxopterygium*, *Spondias*, *Tapirira*, and *Thyrsodium* can be large trees, which produce valuable timber.

Anacardium

Tree	Unbuttressed, but basally swollen. Tree up to 35 m, diameter 60-90 (occasionally up to 150) cm. Bole cylindrical, with a little taper, up to 18 m long.
Description of the wood	Sapwood greyish-white, not always demarcated from the heartwood. Heartwood colour variegated, light brown to yellowish, sometimes pinkish-brown. Texture fine. Lustrous. Grain somewhat roey.
Weight	Specific gravity: 500-650 kg/m^3 (12%).
Shrinkage	From green to ovendry: radial 2.7%, tangential 5.0%, volumetric 9.5%.
Seasoning properties	Moderately difficult to air-season.
Mechanical properties	Crushing strength: 330 kg/sq. cm. Static bending: 835 kg/sq. cm. Modulus of elasticity: 83 kg/sq. cm x 1000.
Working properties	Works easily, finishes smoothly.
Durability	Resistant to termites, but prone to pinhole borer attack.
Uses	Suitable for interior work, and plywood.
Supply	Limited.
Vernacular names	Hunudi, Kawarui, Roroi, Ubudi, Wild cashew (GU). Bosch-kasjoe, Kasjoe (SU). Bouchi kassoun, Moni (FG).

Astronium

Tree	Unbuttressed, but basally swollen. Tree up to 35 m, diameter 40-50 cm. Bole cylindrical, up to 18 m long.
Description of the wood	Sapwood light-brown, sharply demarcated from the rich red to orange-brown, streaked with black heartwood. Texture fine. High satiny lustre. Grain fine, straight to roey.
Weight	Specific gravity: 850-1000 kg/m^3 (12%).
Shrinkage	From green to ovendry: radial 4.5%, tangential 8.6%, volumetric 14.2%.
Working properties	Works easily, turns readily, finishes smoothly, takes a high polish.
Mechanical properties	Crushing strength: 775 kg/sq. cm. Static bending: 1705 kg/sq. cm. Modulus of elasticity: 174 kg/sq. cm x 1000.
Durability	Very durable.
Uses	Suitable for turnery, tool handles, and veneer.
Supply	Limited.
Vernacular names	Bastard-purpleheart, Bauwaua (GU).

Loxopterygium

Tree	Low-buttressed tree. Up to 35 m, diameter 40-70 cm. Bole well-formed, 15-20 m long.
Description of the wood	Sapwood 5-8 cm thick, pale yellow or light grey-brown, often not sharply demarcated from the light-brown to red-brown coloured heartwood, often attractively figured with numerous narrow to rather wide darker stripes and streaks. Radial gum ducts in some rays frequently cause dark oil specks on the tangential surfaces. Texture medium. Lustre medium. Grain straight, sometimes interlocked or wavy.

Weight	Specific gravity: 600-800 kg/m³ (12%).
Seasoning properties	Variable, rated as moderately difficult to air season. Heavier material dries slowly and tends to warp and check.
Mechanical / Physical properties	Air dry, the timber compares closely with Burma Teak in all strength properties, except compression and tension perpendicular to the grain.
Working properties	Saws, planes and turns easily, with good finish. Gluing requires care. The resin/oily gum in the wood may cause some difficulties in varnishing.
Durability	Resistant to decay, moderately resistant to termites.
Preservation	Highly resistant to impregnation with preservatives.
Uses	Panelling, high-grade furniture and cabinet work, doors, plywood.
Supply	Well-distributed, but not in large quantities.
Vernacular names	Aupar, Hububalli, Koika, Kwipari, Kwiparirye (GU). Boesi mahonie, Hoeboeballi, Koeipjarie, Slangenhout, Snekiehoedoe (SU). Koiha, Kooel pialli (FG).

Spondias

Tree	Unbuttressed, basally swollen. Tree up to 40 m, diameter up to 1.20 m. Bole cylindrical, up to 18-24 m long.
Description of the wood	No difference between sapwood and heartwood, colour cream to whitish-yellow. Texture rather coarse. Lustre medium. Grain fairly straight, sometimes slightly interlocked. Contains a sticky resin.

Weight	Specific gravity: 350-600 kg/m³ (12%).
Shrinkage	From green to ovendry: radial 3.8%, tangential 6.4%, volumetric 11.2%.
Seasoning properties	Seasons at a moderate rate with some warping, and end surface checking.
Mechanical properties	Crushing strength: 432 kg/sq. cm. Static bending: 1000 kg/sq. cm. Modulus of elasticity: 84 kg/sq. cm x 1000.
Working properties	Works easily, finishes smoothly.
Durability	Perishable, susceptible to stain and termites.
Preservation	Easy (sapwood) to moderate (heartwood).
Uses	Plywood, interior construction, matches.
Supply	Only available in small quantities.
Vernacular names	Hog plum, Hubu, Mope, Plum, Rup (GU). Hobo, Hoebie, Hoeboe, Monpé, Mopé (SU). Maubin, Mombin, Mopi, Prunier mombin (FG).

Tapirira

Tree	Plank buttresses up to 1.5 m. Tree up to 35 m, diameter up to 60 cm. Bole cylindrical to somewhat flattened, tapering, up to 18 m long.
Description of the wood	Sapwood slightly lighter in colour than heartwood, which is golden light-brown to brown-yellow, sometimes light-pink. Texture fine. Lustrous. Grain straight.
Weight	Specific gravity: 440-560 kg/m³ (12%).
Seasoning properties	Easy to season.

Working properties	Works easily, finishes smoothly, turns and bores satisfactorily.
Durability	Moderately resistant to decay, sapwood very susceptible to blue stain.
Uses	Interior construction, plywood.
Vernacular names	Atapiriri, Duka, Makarin, Matchwood, Warimia (GU). Atapiririe, Danlieba, Doka, Mankrappa, Walimia, Wariennia (SU). Agagnamaïe, Bois à huile, Bois tepiré, Mombin sauvage (FG).

Thyrsodium

Description of the wood	Sapwood not demarcated from heartwood, colour light pink to yellow-brown.
Weight	Specific gravity: *T. guianense*: 710-870 kg/m^3 (12%), *T. spruceanum*: 600-650 kg/m^3 (12%).
Vernacular names	Uluballi (GU). Encens rouge, Grand moni, Moni (FG).

LITERATURE ON WOOD AND TIMBER

Anonymous. Unknown. Guyana Woods. 21 species. Guyana Forestry Commission, Georgetown.

Béna, P. 1960. Essences forestières de Guyane. Bureau Agricole et Forestier Guyanais. Imprimerie Nationale, Paris.

Benoist, R. 1933. Les bois de la Guyane française. Arch. Bot. Mém. 5(1).

Berni, C.A., E. Bolza & F.J. Christensen. 1979. South American timbers – The characteristics, properties and uses of 190 species. C.S.I.R.O., Melbourne.

Chudnoff, M. 1984. Tropical Timbers of the World. Agri-culture Handbook Number 607. U.S. Forest Service, Madison, WI.

Détienne, P., P. Jacquet & A. Mariaux. 1982. Manuel d'identification des bois tropicaux. Tome 3. Guyane française. C.T.F.T., Nogent-sur-Marne.

Détienne, P. & P. Jacquet. 1983. Atlas d'identification des bois de l'Amazonie et des régions voisines. C.T.F.T., Nogent-sur-Marne.

Dong, Z. & P. Baas. 1993. Wood anatomy of trees and shrubs from China. V. Anacardiaceae. IAWA Journl. 14(1): 87-102.

Fanshawe, D.B. 1948. Forest Products of British Guiana. Part 1. Principal Timbers. Forestry Bulletin 1 (New Series).

Gérard, J., R.B. Miller & B.J.H. ter Welle. 1996. Major Timber Trees of Guyana. Tropenbos Series 15. Tropenbos Foundation, Wageningen.

Hess, R.W. 1946. Identification of new world timbers. Pt. II. Anacardiaceae. Trop. Woods 87: 11-34.

Japing, C.H. & H.W. Japing. 1960. Houthandboek Surinaamse houtsoorten. 's Lands Bosbeheer, Paramaribo.

Lindeman, J.C. & A.M.W. Mennega. 1963. Bomenboek voor Suriname. Dienst 's Lands Bosbeheer, Paramaribo.

Mennega, A.M.W. 1948. Surinam Timbers I. General introduction, Guttiferae, Vochysiaceae, Anacardiaceae, Icacinaceae. Nat. Wet. Studiekring v. Suriname en Curaçao. No. 3. Marinus Nijhoff, Den Haag.

Mennega, E.A., W.C.M. Tammens-de Rooij & M.J. Jansen-Jacobs (Eds.). 1988. Checklist of woody plants of Guyana. Tropenbos Technical Series 2. Tropenbos Foundation, Ede.

Metcalfe, C.R. & L. Chalk. 1950. Anatomy of the Dicotyledons. Volume 1. Clarendon Press, Oxford.

Pfeiffer, J.Ph. 1926. De houtsoorten van Suriname. Med. XXII. Afd. Handelsmuseum no. 6. De Bussy, Amsterdam.

Polak, A.M. 1992. Major timber trees of Guyana. Tropenbos Series 2. Tropenbos Foundation, Wageningen.

Record, S.J. 1939. American woods of the family Anacardiaceae. Trop. Woods 60: 11-45.

Record, S.J. & R.W. Hess. 1943. Timbers of the New World. Yale University Press, New Haven, CT.

Terrazas, T. 1994. Wood anatomy of the Anacardiaceae: ecological and phylogenetic interpretation. Ph.D. dissertation, University of North Carolina at Chapel Hill.

Terrazas, T. & W.C. Dickison. (In prep.). Wood anatomy of Anacardiaceae. IAWA Journl.

Vink, A.T. 1965. Surinam timbers. Surinam Forest Service, Paramaribo.

Welle, B.J.H. ter. 1976. Silica grains in woody plants of the Neotropics, especially Surinam. In: Wood structure in biological and technological research (Eds. P. Baas, A.J. Bolton & D.M. Catling). Leiden Bot. Series 3: 107-142.

Wheeler, E.A., P. Baas & P.E. Gasson (Eds.). 1989. IAWA list of microscopic features for hardwood identification. IAWA Bull. n.s. 10(3): 219-332.

NUMERICAL LIST OF ACCEPTED TAXA

1. Anacardium L.
   1-1. A. amapaënse J.D. Mitch.
   1-2. A. fruticosum J.D. Mitch. & S.A. Mori
   1-3. A. giganteum W. Hancock ex Engl.
   1-4. A. occidentale L.
   1-5. A. spruceanum Benth. ex Engl.

2. Astronium Jacq.
   2-1. A. fraxinifolium Schott
   2-2. A. lecointei Ducke
   2-3. A. ulei Mattick

3. Cyrtocarpa Kunth
   3-1. C. velutinifolia (R.S. Cowan) J.D. Mitch. & Daly

4. Loxopterygium Hook. f.
   4-1. L. sagotii Hook. f.

5. Mangifera L.
   5-1. M. indica L.

6. Schinus L.
   6-1. S. terebinthifolia Raddi

7. Spondias L.
   7-1. S. dulcis Parkinson
   7-2. S. mombin L.
   7-3. S. purpurea L.

8. Tapirira Aubl.
   8-1. T. bethanniana J.D. Mitch.
   8-2. T. guianensis Aubl.
   8-3. T. obtusa (Benth.) J.D. Mitch.

9. Thyrsodium Salzm. ex Benth.
   9-1. T. guianense Sagot ex Marchand
   9-2. T. puberulum J.D. Mitch. & Daly
   9-3. T. spruceanum Salzm. ex Benth.

COLLECTIONS STUDIED[5]

GUYANA

Abraham, A.A., 17 (8-2).

Anderson, C.W., 37 = FD 175 (4-1).

Archer, W.A., 2312, 2324 (1-3); 2434 (4-1).

Bartlett, A.W., s.n. (1-3).

Boom, B. et al., 8067 (8-3); 8151, 8163 (8-2); 8195 (9-3); 8242 (8-3).

Carrick, J., 1030 (1-4).

Cook, C.D.K., 163 (3-1).

Cooper, A., 400 (1-4).

Cowan, R.S. (et al.), 39238 (8-2).

Cox, D.I. (with E.A. Hubbard), 82 (4-1).

Cruz, J.S. de la, 1158 (8-2); 1402 (8-3); 1663, 1821, 1828 (8-2); 1951 (7-2); 1997 (8-2); 2157 (1-4); 2225, 2385, 2438 (8-2); 2551, 2708 (1-4); 2728 (8-2); 2799 (1-4); 2829 (8-2); 3341 (1-4); 3342, 3385, 3590 (5-1); 3648 (1-4); 3834 (8-2); 3839 (7-2); 3948, 4253, 4283 (8-2); 4324 (1-4); 4408, 4436 (8-2); 4504 (1-3); 4538 (8-2).

Dahlgren, B.E. (with A.C. Persaud), s.n. (1-4).

Davis, D.H., 849 (1-4).

Davis, T.A.W., 109 = FD 2100 (4-1); 155 = FD 2146, 165 = FD 2156 (2-3); 558 = FD 2596 (4-1).

Fanshawe, D.B., 30 = FD 2598 (1-3); 544 = FD 3280, 930 = FD 3666, 947 = FD 3683 (7-2); 1492 = FD 4228 (8-2); 1557 = FD 4293 (4-1); 1606 = FD 4342 (7-2); 2259 = FD 4995 (1-3); 2732 = *FD 5531* (8-2); 3521 = FD 7117 (1-4).

FD (= Forest Department British Guiana see also Anderson, Fanshawe, Hohenkerk, Lockie and Wilson-Browne), 175A, 175 (4-1); 869 (7-2); *2019* (9-3); 2100 (4-1); 2146, 2156 (2-3); 2596 (4-1); 2598 (1-3); 3280, 3666, 3683 (7-2); 4228 (8-2); 4293 (4-1); 4342 (7-2); 4995 (1-3); *5531* (8-2); *5650* (3-1); 5733 (8-2); 5919 (3-1); 7117 (1-4).

Gillespie, L.J. (et al.), 740, 2505, 2509, 3016, 4254 (8-2).

Gleason, H.A., 293 (8-2); 382 (1-4); 765 (8-2).

Goodland, R., 858 (1-4); 1046 (1-3).

Hardy, T., 368 (1-4).

Harris, S.A., 434 (8-2).

Harrison, S.G. (et al.), 680 (8-2); 738 (1-4); 1006 (8-2); 1679 (5-1).

Hitchcock, A.S., 17171 (1-4); 17237 (8-2).

Hoffman, B. (et al.), 716 (1-4); 794, 897, 905 (8-2); 940 (1-4); 1071, 1216 (3-1); 1603 (1-2); 2560 (8-2); 3416 (1-2).

Hohenkerk, L.S., D13 = FD 869 (7-2); C26 = FD 175A (4-1).

Im Thurn, E.F., s.n. (8-2).

Jansen-Jacobs, M.J. et al., 734, 750 (7-2); 1055 (8-2); 1202 (7-2); 1345, 1959 (8-2); 2120 (3-1); 2565 (1-3); 2648 (7-2); 3099, 3141 (8-2); 3316 (7-2); 3559, 3728 (3-1).

[5] Numbers in italics represent type collections.

Jenman, G.S., s.n. (8-2); 108 (7-2); 452, 457 (8-2); 1109 (1-4); 1923, 2354 (8-3); 2437 (8-2); 2462 (8-3); 3578 (7-2); 3905 (8-2); 4094 (4-1); 4335 (1-3); 4550 (1-4); 4684, 4776 (8-2); 4849 (4-1); 4875 (1-3); 5182 (4-1); 6482 (7-2); 6623 (8-2).

Joseph, G.F., s.n. (1-4).

Kelloff, C. et al., 632, 635 (8-2).

Lang, H. & A.C. Persaud, 50, 149, 357 (1-4).

Lee, G. & R. Persaud, 4 (1-4).

Leechman, A., s.n. (7-2).

Lockie, J.R., s.n. = *FD 2019* (9-3).

Maas, P.J.M. (et al.), 3480, 3481, 3578 (8-2).

Maguire, B. (et al.), *32304*, 43865 (1-2).

McDowell, T. (et al.), 1740, 2032 (8-2).

Omawale & R. Persaud, 62 (7-2); 78 (1-4); 94 (7-1); 115, 116 (1-4).

Persaud, A.C., 6, 10 (8-2); 18 (1-4); 35 (8-3); 36 (1-3).

Peterson, P.M. (et al.), 7464 (8-2).

Pinkus, A.S., 192 (8-2).

Pipoly, J.J. (et al.), 7346 (6-1); 7434, 7485 (8-2); 7684 (1-2); 8249 (7-2); 8334, 8431, 9146 (8-2); 9162 (1-4); 9167, 9224 (8-2); 9270 (1-4); 9333 (8-2); 9336, 9395 (1-4); 9449 (8-2); 9497 (1-4); 9560, 9620 (8-2); 11559, 11569 (1-4); 11584 (8-2); 11742 (7-2); 11759 (8-2); 11761 (1-4); 11782 (8-2).

Polak, M. (et al.), 157, 193 (8-2); 295 (4-1); 303 (2-3); 309 (4-1).

Reinders, M.A. (et al.), 23 (1-4); 81 (7-2).

Sandwith, N.Y., 509, 560 (8-3).

Schomburgk, R. (Rich. or Rob.), s.n. (8-2); 728 (4-1).

Schomburgk, Rich., 728, 755, 1340 = Rob. II 789, 1406 = Rob. II 793 (8-2); *1442 = Rob. II 892* (8-3); 1482 = Rob. II 916, 1483 = Rob. II 915 (8-2); 1525 = Rob. II 919 (7-2); 1706 = Rob. II 1010 (8-2).

Schomburgk, Rob., I 174 (8-2); *I 892* (9-3); II 37 (7-2); II 789 = Rich. 1340, II 793 = Rich. 1406 (8-2); *II 892 = Rich. 1442* (8-3); II 915 = Rich. 1483, II 916 = Rich. 1482 (8-2); II 919 = Rich. 1525 (7-2); II 1010 = Rich. 1706 (8-2).

Smith, A.C., 2437 (8-2); 3103 (7-2); 3126 (2-3); 3185 (9-3); 3438 (7-2); 3466 (2-3); 3512 (4-1); 3609 (9-3).

Steege, H. ter, 378 (8-2).

Stevens, F.L., s.n. (8-2).

Tillett, S.S. & C.L. Tillett, 45800, 45831, (8-2).

Tutin, T.G., 189 (8-2).

Warren, G.B., s.n. (1-4).

Wilson-Browne, G., *112* = *FD 5650* (3-1); 304 = FD 5733 (8-2); 518 = FD 5919 (3-1).

SURINAME

BBS (Bosbeheer Suriname), 208 (1-3); 1131 (7-2); 1143 (1-3).

Berthoud-Coulon, M., 548 (7-2).

Boldingh, I. 3882? (1-4).

Borsboom, N.W.J. in LBB 12005 (9-2).

BW (Boschwezen), 71 (4-1); 76 (8-2); 81 (4-1); 168 (7-2); 268 (1-4); 278 (1-3); 331, 443 (8-2);

613 (= tree no. SO 526), 844 (1-4); 1237 (= tree no. SO 552) (8-2); 1256 (7-2); 1359 (1-3); 1611 (= tree no. SO 552) (8-2); 1736 (= tree no. BB 1136), 2429 (4-1); 2511 (= tree no. WM 1664) (7-2); 2558 (= tree no. SO 526) (1-4); 2739 (= tree no. SO 552) (8-2); 2800 (= tree no. WM 1664) (7-2); 4021 (= tree no. WM 1665) (8-2); 4025 (= tree no. WM 1664) (7-2); 4089 (= tree no. Z I IV) (4-1); 4188 (= tree no. SO 102); 4316, 4323 (1-4); 4480 (= tree no. WM 1664) (7-2); 4518 (= tree no. Z I IV) (4-1); 4542 (= tree no. WM 1665) (8-2); 4697 (= tree no. Z I IV) (4-1); 4930 (1-3); 4970, 5007 (= tree no. WM 1664) (7-2); 5051 (= tree no. Z I IV) (4-1); 5356 (= tree no. BB 1034) (8-2); 5466 (= tree no. SO 526) (1-4); 5512, 5555 (= tree no. SO 526) (1-4); 5572 (= tree no. WM 1665) (8-2); 5760, 5821 (= tree no. Z I IV), 5866 (= tree no. SO 833) (4-1); 6136 (= tree no. SO 102) (1-3); 6231 (= tree no. BB 1034) (8-2); 6387 (7-2); 6389 (1-4); 6590 (= tree no. BB 1136) (4-1); 6622 (1-5); 6877 (= tree no. BB 1034) (8-2).

Burger, D. (with E.M.C. Helstone), 16 (4-1).

Collector indigenous, s.n. (1-4); s.n., 181 (5-1); 2018 (4-1).

Daniëls, A.G.H. (with F.P. Jonker), 908 (4-1).

Donselaar, J. van, 1905 (9-1); 2041 (1-5).

Dumontier, F.A.C., s.n. (4-1).

Elburg, J. in LBB 9354 (2-2); 99879 (8-2); 13380 (2-2).

Florschütz, J.& P.A., 1634 (4-1).

Focke, H.C., s.n. (1-4); 919 (5-1).

Glocker, E.F. von, s.n. (5-1).

Gonggrijp, J.W., BW 54 (1-4).

Hekking, W.H.A., 820 (8-2).

Helstone, E.M.C., 227 = LBB 8092 (2-3).

Heijligers, P.C. (et al.), 39, 333 (1-4).

Hostmann, F.W.R. (et al.), 368, 604A, 853 (8-2).

Im Thurn, E.F., s.n. (8-2).

Jansma, R. in LBB 15908 (1-4).

Jiménez-Saa, J.H., 1500 = LBB 14241 (1-3); 1580 = LBB 14313 (8-2); 1648 = LBB 14381 (1-5).

Jonker-Verhoef, A.M.E. & F.P. Jonker, 463 (1-4).

Kramer, K.U. & W.H.A. Hekking, 2496 (1-4); 3140 (5-1).

Kuyper, J., 32 (1-4).

Lanjouw, J. & J.C. Lindeman, 322 (5-1); 403 (9-1); 459 (8-2); 998 (7-2); 1083 (8-2); 1388 (7-2); 2307, 2496 (9-1); 3152 (7-2).

LBB (Suriname, see also Borsboom, Elburg, Helstone, Jansma, Jiménez-Saa, Maas & Tawjoeran), 8092 (2-3); 9354 (2-2); 9403 (8-3); 9879 (8-2); 10734 (2-3); 10735, 10820 (1-5), 12005 (9-2); 13380 (2-2); 14241 (1-3); 14313 (8-2); 14381 (1-5), 14659 (9-1); 15908 (1-4).

Lems, K., 5081 (8-2); 5083 (4-1); 5105 (1-4).

Lindeman, J.C., 4773, 4876 (9-3); 5841 (4-1); 5920, 5929 (9-1); 6057 (4-1); 6080 (2-3); 6794 (9-3); 6804 (1-5); 6856 (2-2).

Lindeman, J.C. & A.R.A. Görts et al., 118 (8-2).

Lindeman, J.C. & A.C. de Roon, 7143 (9-3).

Maas, P.J.M. & J.A. Tawjoeran in LBB 10734 (2-3); 10735, 10820 (1-5).

Maguire, B. (et al.), 22743 (1-4); 24603, 24631, 24953 (8-2).

Mori, S.A. & A. Bolten, 8280, 8418, 8679 (8-2).

Oldenburger, F.H.F. et al., 327 (8-2); 496 (2-1); 590 (1-3); 1279 (9-3).

Procter, J., 4761 (4-1).

Rombouts, H.E., 765 (1-4).

Samuels, J.A., 120 (1-4); 190, 215 (8-2).

Schimper, s.n. 368 (8-2).

Schulz, J.P., 7264 (8-2); 7531 (2-2); 7958 (9-3).

Soeprato, 10 (1-4).

Splitgerber, F.L., s.n. (8-2).

Stahel, G., Wilhelmina Exp., 59 (1-3); 72 (4-1).

Stahel, G., Woodherbarium, 75, 76 (8-2); 81 (4-1); 168 (7-2); 268 (1-4); 278 (1-3).

Tawjoeran, J.W. in LBB 14659 (9-1).

Tresling, J., 159 (1-4); 393 (1-3).

Versteeg, G.M., 129 (1-4).

Went, F.A.F.C., 224, 466 (1-4).

Wildschut, J.T. & P.A. Teunissen in LBB 11626 (1-4).

Wullschlägel, H.R., 930 (7-2); 1675 (8-2).

FRENCH GUIANA

Aublet, J.B.C.F. d', s.n. (8-2).

Aubréville, A., s.n. (8-2); 90 (7-2); 207 (8-2).

Barrier, S., 4182 (8-2); 4946 (9-2); 4994 (8-1); 5187 (7-2).

Béna, P.C. (see Service Forestier Guyane Française), 1027, 1320 (9-2).

Benoist, R., s.n. (7-2); 9, 139, 201 (8-2); 1237 (4-1); 1714 (8-4).

Billiet, F. et al., 1435 (8-2); 1442 (1-4).

Boom, B. et al., 1750 (8-2); 2006 (9-2); 2223 (8-3); 2439 (9-3); 2465 (8-2).

Bordenave, B., 130, 147 (8-2); 173 (7-2); 463 (8-3).

Broadway, W.E., 273 (1-4); 550, 782 (8-2); 973 (1-4).

CAY, 39 (9-2).

Cowan, R. S. (et al.), 38025 (8-2); 38029 (1-4); 38958 (8-2).

Cremers, G., 5104, 5108 (8-2); 7781, 8508 (7-2).

Feuillet, C., 1418 (9-1); 2970 (8-2).

Fleury, M., 222 (8-2); 729 (7-3); 874 (5-1); 875 (1-4).

Gandoger, M., 96, 119 (8-2).

Geay, M.F., 916 (1-4).

Gentry, A. et al., 62977 (8-3).

Granville, J.J. de (et al.), BC.55 (9-1); BC.94 (8-2); 339 (9-1); 1354 (8-2); 2994 (7-3); 3392 (9-1); 5002, 5158, 52644 (8-2); B.5386 (8-3); B.5398 (8-1); *B.5486* (9-2); 5630 (8-1); 5766 (7-2); 6627 (8-2); 9498 (4-1).

Grenand, P., 1422 (9-2).

Hallé, F., 611 (1-5); 1047 (8-2); 2840 (8-1).

Hoff, M. (et al.), 5082 (8-2); 5089 (1-4); 5144, 5351, 5358 (8-2); 6274 (8-3).

Kodjoed-Bonneton, J.F., 144 (8-2).

Larpin, D., 693 (8-2).

Leblond, J.B., 283 (1-4); 502, 1792 (8-2).

Lemée, A.M.V., s.n. (1-4), (8-2), (7-2), 1525 (8-2).

Lemoine, M., 7810 (8-2); 7839 (1-4).

Leprieur, F.R.M., s.n. (8-2), (1-4), (8-2).

Leguillou, E.J.F., s.n. (1-4), (8-2).

Lescure, J.P., 397 (9-2), 1975 (7-2).

Loubry, D., 1247 (8-1); 1750 (1-5).

Maas, P.J.M. et al., 2206 (8-2).

Maire, s.n. (8-2).

Martin, J., s.n. (8-2), (8-2), (8-3); 43 (8-2).

Mélinon, E., s.n. (1-4), (8-3), (1-5), (8-2), (83); 18 (8-2); 18 (8-3); 36 (8-2); 44 (1-5); 48 (1-4); 69 (4-1); 71, 86, 125 (8-2); 179, 187, 298 (1-4).

Mennega, A.M.W. (et al.), 868, 869 (4-1).

Mori, S.A. (et al.), 8959 (8-3); 15156, *15253*, 15335 (8-1); 15684 (1-5); 15688 (8-3); 18752 (8-1); 18766 (8-2); 18775 (8-3); 18777 (7-3); 18778 (5-1); 21534 (7-2); 21588 (8-3); 21687 (9-1); 22857 (8-3); 23435, 23642 (9-1).

Oldeman, R.A.A. (et al.), 215 (8-2); 242 (9-2); B.729, 844 A, 1076, B.1294 (8-2); B.1318 (8-3); 1336 (8-2); B.1393 (8-3); B.1647, 1689, 1758, B.1921 (8-2); 2582 (9-2); 2640 (8-2); 3244 (9-1); 3257 (8-3).

Parker, C.S., s.n. (8-2).

Perrottet, G.S., s.n. (8-2).

Poiteau, P.A., s.n. (8-2).

Prévost, M.F., 618, 827, 1092 (8-2); 2161 (2-3).

Puig, H., 32 (8-2).

Richard, L.C., s.n. (1-4).

Riera, B.J.J.L. (et al.), 39 (9-2); 436 (9-1); 578 (8-2); 579, 912 (8-3); 938 (7-3); 1073 (9-2); B.1240 (8-3); 1472 (8-2); 1892 (8-1).

Sabatier, D. (et al.), 74 (9-1); 617, 714 (8-1); 889 (1-5); 1244 (9-2); 1404 (9-1); 1710 (8-1); 1736 (9-2); 1811 (8-2); 1874 (9-2); 1882, 1933 (2-3); 1955 (9-1); 1980 (2-3); 2046 (8-1); 2357 (9-2); 2362 (8-1); 2486, 2521 (1-5); 2613 (9-2); 2838 (9-3); 2853 (2-3); 3026 (8-3); 3027 (8-2); 3146 (1-1); 3577 (8-1); 3706 (9-3); 3723 (1-5); 3813 (8-3); 3814 (8-1).

Sagot, P.A., s.n. (1-4), (8-2), (7-1), (7-3); 194 (1-4); 196 (7-1); 197 (8-2); 197 (7-2); *973* (4-1); *1202* (9-1).

Sastre, C. (et al.), 215 (1-4); 5992, 6344 (8-2).

Schnell, R., 12090 (8-2).

Service Forestier Guyane Française (= BAFOG), 13N (9-2); 71M (9-1); 94M (8-2); 114M (9-2); 193M (1-5); 198M (4-1); 202M (7-2); 213M (9-2); 257M, 1027 (9-1); 3195, 4216 (1-4); 4293 (8-3); 7053 (8-2); 7061, 7090 (8-3); 7091 (8-2); 7099 (9-2); 7162, 7223 (8-2); 7297 (9-2); 7305, 7311 (8-2); 7575 (4-1); 7645 (1-5); 7713 (1-5); 7839 (1-4).

Taverne, B., 51 (7-3).

Vaillant, s.n., (1-4).

Viellescazes, A., 476 (9-2).

Villers, J.F. (et al.), 1553, 1835 (9-2); 2242 (8-1).

Wachenheim, H., s.n., 33, 169 (1-4); 216 (8-3); 236 (9-1); 237 (8-2).

INDEX TO SYNONYMS AND TYPE SPECIES

INDEX TO VERNACULAR AND TRADE NAMES

rub 7
savanna weti-oedoe 8-2
slangenhout 4-1
sneki-oedoe 4-1
spanish plum 7-3
taite tukasina 9-2
tapeliwa 7-2
tapereba 7-2
tapiriri 8-2
tatapilili 8-2
tatapirica 8-2

tingimoni 9-2
ubudi 1-3
uluballi 9
walimia 8
wariennia 8
warimia 8-2
weti-oedoe 8-2
wild cashew 1
witte doeka 8-2
zwarte doeka 8-2

Alphabetic list of families of series A occurring in the Guianas

Defined as in Cronquist, 1981, and numbered in his sequence, with alternative names. Those published, with chronological fascicle number and year.

Abolbodaceae			Campanulaceae	162	
(see Xyridaceae	182)	15. 1994	(incl. Lobeliaceae)		
Acanthaceae	156		Cannaceae	195	1. 1985
(incl. Thunbergiaceae)			Canellaceae	004	
(excl. Mendonciaceae	159)		Capparaceae	067	
Achatocarpaceae	028		Caprifoliaceae	164	
Agavaceae	202		Caricaceae	063	
Aizoaceae	030		Caryocaraceae	042	
(excl. Molluginaceae	036)		Caryophyllaceae	037	
Alismataceae	168		Casuarinaceae	026	11. 1992
Amaranthaceae	033		Cecropiaceae	022	11. 1992
Amaryllidaceae			Celastraceae	109	
(see Liliaceae	199)		Ceratophyllaceae	014	
Anacardiaceae	129	19. 1997	Chenopodiaceae	032	
Anisophylleaceae	082		Chloranthaceae	008	
Annonaceae	002		Chrysobalanaceae	085	2. 1986
Apiaceae	137		Clethraceae	072	
Apocynaceae	140		Clusiaceae	047	
Aquifoliaceae	111		(incl. Hypericaceae)		
Araceae	178		Cochlospermaceae		
Araliaceae	136		(see Bixaceae	059)	
Arecaceae	175		Combretaceae	100	
Aristolochiaceae	010		Commelinaceae	180	
Asclepiadaceae	141		Compositae		
Asteraceae	166		(= Asteraceae	166)	
Avicenniaceae			Connaraceae	081	
(see Verbenaceae	148)	4. 1988	Convolvulaceae	143	
Balanophoraceae	107	14. 1993	(excl. Cuscutaceae	144)	
Basellaceae	035		Costaceae	194	1. 1985
Bataceae	070		Crassulaceae	083	
Begoniaceae	065		Cruciferae		
Berberidaceae	016		(= Brassicaceae	068)	
Bignoniaceae	158		Cucurbitaceae	064	
Bixaceae	059		Cunoniaceae	081a	
(incl. Cochlospermaceae)			Cuscutaceae	144	
Bombacaceae	051		Cycadaceae	208	9. 1991
Bonnetiaceae			Cyclanthaceae	176	
(see Theaceae	043)		Cyperaceae	186	
Boraginaceae	147		Cyrillaceae	071	
Brassicaceae	068		Dichapetalaceae	113	
Bromeliaceae	189	p.p. 3. 1987	Dilleniaceae	040	
Burmanniaceae	206	6. 1989	Dioscoreaceae	205	
Burseraceae	128		Dipterocarpaceae	041a	17. 1995
Butomaceae			Droseraceae	055	
(see Limnocharitaceae	167)		Ebenaceae	075	
Byttneriaceae			Elaeocarpaceae	048	
(see Sterculiaceae	050)		Elatinaceae	046	
Cabombaceae	013		Eremolepidaceae	105a	
Cactaceae	031	18. 1997	Ericaceae	073	
Caesalpiniaceae	088	p.p. 7. 1989	Eriocaulaceae	184	
Callitrichaceae	150		Erythroxylaceae	118	

Euphorbiaceae	115	Loranthaceae	105
Euphroniaceae	123a	(excl. Viscaceae	106)
Fabaceae	089	Lythraceae	094
Flacourtiaceae	056	Malpighiaceae	122
(excl. Lacistemaceae	057)	Malvaceae	052
(excl. Peridiscaceae	058)	Marantaceae	196
Gentianaceae	139	Marcgraviaceae	044
Gesneriaceae	155	Martyniaceae	
Gnetaceae	209	9. 1991	(see Pedaliaceae | 157)
Gramineae		Mayacaceae	183
(= Poaceae	187)	8. 1990	Melastomataceae	099	13. 1993
Gunneraceae	093	Meliaceae	131
Guttiferae		Mendonciaceae	159
(= Clusiaceae	047)	Menispermaceae	017
Haemodoraceae	198	15. 1994	Menyanthaceae	145
Haloragaceae	092	Mimosaceae	087
Heliconiaceae	191	1. 1985	Molluginaceae	036
Henriquesiaceae		Monimiaceae	005
(see Rubiaceae | 163)	Moraceae	021	11. 1992
Hernandiaceae	007	Moringaceae	069
Hippocrateaceae	110	16. 1994	Musaceae	192	1. 1985
Humiriaceae	119	(excl. Strelitziaceae	190)
Hydrocharitaceae	169	(excl. Heliconiaceae	191)
Hydrophyllaceae	146	Myoporaceae	154
Icacinaceae	112	16. 1994	Myricaceae	025
Hypericaceae		Myristicaceae	003
(see Clusiaceae	047)	Myrsinaceae	080
Iridaceae	200	Myrtaceae	096
Ixonanthaceae	120	Najadaceae	173
Juglandaceae	024	Nelumbonaceae	011
Juncaginaceae	170	Nyctaginaceae	029
Krameriaceae	126	Nymphaeaceae	012
Labiatae		(excl. Nelumbonaceae	010)
(= Lamiaceae	149)	(excl. Cabombaceae	013)
Lacistemaceae	057	Ochnaceae	041
Lamiaceae	149	Olacaceae	102	14. 1993
Lauraceae	006	Oleaceae	152
Lecythidaceae	053	12. 1993	Onagraceae	098	10. 1991
Leguminosae		Opiliaceae	103	14. 1993
(= Mimosaceae	087)	Orchidaceae	207
+ Caesalpiniaceae	088)	p.p. 7. 1989	Oxalidaceae	134
+ Fabaceae	089)	Palmae	
Lemnaceae	179	(= Arecaceae	175)
Lentibulariaceae	160	Pandanaceae	177
Lepidobotryaceae	134a	Papaveraceae	019
Liliaceae	199	Papilionaceae	
(incl. Amaryllidaceae)		(= Fabaceae	089)
(excl. Agavaceae	202)	Passifloraceae	062
(excl. Smilacaceae	204)	Pedaliaceae	157
Limnocharitaceae	167	(incl. Martyniaceae)	
(incl. Butomaceae)		Peridiscaceae	058
Linaceae	121	Phytolaccaceae	027
Lissocarpaceae	077	Pinaceae	210	9. 1991
Loasaceae	066	Piperaceae	009
Lobeliaceae		Plantaginaceae	151
(see Campanulaceae	162)	Plumbaginaceae	039
Loganiaceae	138	Poaceae	187	8. 1990

Podocarpaceae	211	9. 1991	Symplocaceae	078	
Podostemaceae	091		Taccaceae	203	
Polygalaceae	125		Tepuianthaceae	114	
Polygonaceae	038		Theaceae	043	
Pontederiaceae	197	15. 1994	(incl. Bonnetiaceae)		
Portulacaceae	034		Theophrastaceae	079	
Potamogetonaceae	171		Thunbergiaceae		
Proteaceae	090		(see Acanthaceae	156)	
Punicaceae	097		Thurniaceae	185	
Quiinaceae	045		Thymeleaceae	095	
Rafflesiaceae	108		Tiliaceae	049	17. 1995
Ranunculaceae	015		Trigoniaceae	124	
Rapateaceae	181		Triuridaceae	174	5. 1989
Rhabdodendraceae	086		Tropaeolaceae	135	
Rhamnaceae	116		Turneraceae	061	
Rhizophoraceae	101		Typhaceae	188	
Rosaceae	084		Ulmaceae	020	11. 1992
Rubiaceae	163		Umbelliferae		
(incl. Henriquesiaceae)			(= Apiaceae	137)	
Ruppiaceae	172		Urticaceae	023	11. 1992
Rutaceae	132		Valerianaceae	165	
Sabiaceae	018		Velloziaceae	201	
Santalaceae	104		Verbenaceae	148	4. 1988
Sapindaceae	127		(incl. Avicenniaceae)		
Sapotaceae	074		Violaceae	060	
Sarraceniaceae	054		Viscaceae	106	
Scrophulariaceae	153		Vitaceae	117	
Simaroubaceae	130		Vochysiaceae	123	
Smilacaceae	204		Winteraceae	001	
Solanaceae	142		Xyridaceae	182	15. 1994
Sphenocleaceae	161		(incl. Albolbodaceae)		
Sterculiaceae	050		Zamiaceae	208a	9. 1991
(incl. Byttneriaceae)			Zingiberaceae	193	1. 1985
Strelitziaceae	190	1. 1985	(excl. Costaceae	194)	
Styracaceae	076		Zygophyllaceae	133	
Suraniaceae	086a				

11

10

9

8

7

6

5

4

3

2

1